VOLUME ONE HUNDRED AND SEVENTY SIX

PROGRESS IN MOLECULAR BIOLOGY AND TRANSLATIONAL SCIENCE

The Microbiome

VOLUME ONE HUNDRED AND SEVENTY SIX

PROGRESS IN MOLECULAR BIOLOGY AND TRANSLATIONAL SCIENCE

The Microbiome

Edited by

LORA J. KASSELMAN
*NYU Long Island School of Medicine,
Mineola, NY, United States*

ELSEVIER

ACADEMIC PRESS
An imprint of Elsevier

Academic Press is an imprint of Elsevier
50 Hampshire Street, 5th Floor, Cambridge, MA 02139, United States
525 B Street, Suite 1650, San Diego, CA 92101, United States
The Boulevard, Langford Lane, Kidlington, Oxford OX5 1GB, United Kingdom
125 London Wall, London EC2Y 5AS, United Kingdom

First edition 2020

Notices
Knowledge and best practice in this field are constantly changing. As new research and experience broaden our understanding, changes in research methods, professional practices, or medical treatment may become necessary.

Practitioners and researchers must always rely on their own experience and knowledge in evaluating and using any information, methods, compounds, or experiments described herein. In using such information or methods they should be mindful of their own safety and the safety of others, including parties for whom they have a professional responsibility.

To the fullest extent of the law, neither the Publisher nor the authors, contributors, or editors, assume any liability for any injury and/or damage to persons or property as a matter of products liability, negligence or otherwise, or from any use or operation of any methods, products, instructions, or ideas contained in the material herein.

ISBN: 978-0-12-820797-0
ISSN: 1877-1173

For information on all Academic Press publications
visit our website at https://www.elsevier.com/books-and-journals

Publisher: Zoe Kruze
Acquisitions Editor: Ashlie M. Jackman
Editorial Project Manager: Leticia Lima
Production Project Manager: James Selvam
Cover Designer: Greg Harris

Typeset by SPi Global, India

Contents

Contributors

Andreina Baj
Department of Medicine and Surgery, University of Insubria, Varese, Italy

Sugato Banerjee
Department of Pharmacology and Toxicology, National Institute of Pharmaceutical Education and Research, Kolkata, India

Davide Banfi
Department of Medicine and Surgery, University of Insubria, Varese, Italy

Atin Basuchoudhary
Department of Economics and Business, Virginia Military Institute, Lexington, VA, United States

Michela Bistoletti
Department of Medicine and Surgery, University of Insubria, Varese, Italy

Annalisa Bosi
Department of Medicine and Surgery, University of Insubria, Varese, Italy

Eugene B. Chang
Department of Medicine, Knapp Center for Biomedical Discovery, The University of Chicago, Chicago, IL, United States

Kaberi Chatterjee
Department of Pharmaceutical Sciences and Technology, Birla Institute of Technology, Mesra, Ranchi, India

John A. David
Department of Applied Mathematics, Virginia Military Institute, Lexington, VA, United States

Cristina Giaroni
Department of Medicine and Surgery, University of Insubria, Varese, Italy

Pamela Herd
McCourt School of Public Policy, Georgetown University, Washington, DC, United States

Lindsay M. Hopson
Department of Biochemistry and Molecular Medicine; The McCormick Genomic and Proteomic Center, The George Washington University, Washington, DC, United States

Lora J. Kasselman
NYU Long Island School of Medicine, Mineola, NY, United States

Megan S. Kennedy
Medical Scientist Training Program, Pritzker School of Medicine; Department of Ecology & Evolution, The University of Chicago, Chicago, IL, United States

Pavel Klein
Mid-Atlantic Epilepsy and Sleep Center, Bethesda, MD, United States

Eric Lam
Nassau University Medical Center, East Meadow, NY, United States

Raja Mazumder
Department of Biochemistry and Molecular Medicine; The McCormick Genomic and Proteomic Center, The George Washington University, Washington, DC, United States

Stefanie Prast-Nielsen
Center for Translational Microbiome Research (CTMR), Department of Microbiology, Tumor and Cell Biology, Karolinska Institutet, Stockholm, Sweden

Allison B. Reiss
NYU Winthrop Hospital; NYU Long Island School of Medicine, Mineola, NY, United States

Sabyasachi Sen
Department of Biochemistry and Molecular Medicine; Department of Medicine, The George Washington University, Washington, DC, United States

Neal Shah
NYU Winthrop Hospital, Mineola, NY, United States

Stephanie S. Singleton
Department of Biochemistry and Molecular Medicine, The George Washington University, Washington, DC, United States

Nicholas A. Vernice
Weill Cornell Medicine, New York, NY, United States

Preface

The idea that the gut and the brain are interconnected has been around for centuries. Medical treatments attempting to normalize gut bacteria existed in Ancient Greece, in 4th century Chinese medicine, and in Europe in the 16th and 17th centuries. More recently, the enteric nervous system has been dubbed "the second" brain, underlying the importance of the gut-brain connection. Over the past decade, there has been an explosion of research in this area, with recent emphasis on elucidating the link between the gut microbiome and psychocognitive traits such as personality, mood, and cognition, as well as other nervous system disorders. This is a timely topic to be included in the series *Progress in Molecular Biology and Translational Science*, especially since mood disorders and cognitive dysfunction, e.g., major depression or Alzheimer's disease, continue to be enormous burdens on patients, their families, and society at large. Unfortunately, decades of research into a cure for, or delay of, cognitive decline, such as that seen in Alzheimer's disease, has resulted in few treatment options, underlying the importance of identifying novel mechanisms, such as the gut microbiome, for preserving cognitive function or delaying or reversing cognitive decline. In this volume on The Microbiome, Dr. Chang and her team discuss the assembly of microbial communities across the human body, with special emphasis on the distribution of microbes throughout the intestinal tract. Dr. Giaroni and her team describe the chief endocrine, immune, and neuronal pathways involved in bidirectional communication between the gastrointestinal tract and the brain. Dr. Banerjee and his colleague discuss the role of gut bacteria in motor neuron diseases like multiple sclerosis, Parkinson's disease, and amyotrophic lateral sclerosis. My chapter focuses on the link between the gut microbiome and personality, mood, and cognition in both health and disease. Finally, Dr. Mazumder and his team describe an interdisciplinary approach to gut microbiome research with the goal of mitigating the hindrances of translating findings into a clinical setting. The compilation of these chapters provides a foundation on the gut-brain axis, from which researchers can identify gaps in knowledge, and then conduct appropriate studies to address these gaps. One pressing area is the need for more human studies on the use of probiotics or microbiome-modifying procedures in diseases of the nervous system. It is my hope that well-designed human studies will be conducted in the future to address the potential of working with our own gut microbiome to cure a host of debilitating psychocognitive and other neurologic impairments. I would like to thank all the authors and editors for their hard work in bringing this volume to completion. In addition, I would like to thank my family for putting up with me while this project loomed over me for the last 2 years. Lastly, I hope you enjoy this volume on the microbiome and that it acts as an inspiration for your own work, whether or not it is related to the microbiome!

Lora J. Kasselman, PhD, MPH, Assistant Professor
Foundations of Medicine Department
NYU Long Island School of Medicine
Mineola, NY 11501, United States
2020

CHAPTER ONE

The microbiome: Composition and locations

Megan S. Kennedy[a,b] and Eugene B. Chang[c,*]
[a]Medical Scientist Training Program, Pritzker School of Medicine, The University of Chicago, Chicago, IL, United States
[b]Department of Ecology & Evolution, The University of Chicago, Chicago, IL, United States
[c]Department of Medicine, Knapp Center for Biomedical Discovery, The University of Chicago, Chicago, IL, United States
*Corresponding author: e-mail address: echang@medicine.bsd.uchicago.edu

Contents

Progress in Molecular Biology and Translational Science, Volume 176
ISSN 1877-1173
https://doi.org/10.1016/bs.pmbts.2020.08.013

Abstract

The human body is home to a diverse and functionally important assemblage of symbiotic microbes that varies predictably over different spatial scales, both within and across body sites. The composition of these spatially distinct microbial consortia can be impacted by a variety of stochastic and deterministic forces, including dispersal from different source communities, and selection by regionally-specific host processes for the enrichment of physiologically significant taxa. In this chapter, we review the composition, function, and assembly of the healthy human gastrointestinal, skin, vaginal, and respiratory microbiomes, with special emphasis on the regional distribution of microbes throughout the gastrointestinal tract.

1. Introduction

The gradual discovery that trillions of symbiotic microbes not only occupy sites across the human body, but are often integral players in the healthy physiology of those sites, has been one of the most significant advances in biomedicine since germ theory first revolutionized the field in the 19th century. Indeed, we have likely been evolving and coevolving with our microbes since well before the dawn of our species, and the profound interdependence between host and microbe has thrown into question the very notion of human selfhood. Some would even argue that we must reconceive of human beings (and, for that matter, most multicellular life) as "holobionts," or superorganisms whose evolutionary trajectory can only be considered in light of the composite phenotype conferred by both host and microbial genomes.

Especially over the past decade, scientists have amassed extensive evidence that many aspects of healthy bodily function are intimately dependent upon the particular assemblage of microbes that we carry and their respective genomes, or our "microbiome." It therefore becomes extremely clinically relevant to understand which microbes appear where in the body, and what they are doing. This has been the focus of much foundational exploratory groundwork in the microbiome field, and particularly with the advent of culture-independent methods of bacterial quantification, we have tremendously expanded our knowledge of microbiome composition, function, stability, and resilience in health and disease, across a variety of body sites. In this chapter, we will review the current state of knowledge about the composition and function of the healthy human gastrointestinal, skin, vaginal, and respiratory microbiomes, with particular emphasis on regional specificity of microbes within the gastrointestinal tract, and the ecological processes that may contribute to community assembly in each of these habitats.

2. Ecological principles for understanding community assembly

2.1 Overview

As the microbiome field has matured, researchers have advanced beyond correlative investigations of which microbes appear under what circumstances and begun to pick apart the mechanisms for how these communities assemble, with an eye toward manipulating those processes and engineering the microbiome for therapeutic gain. Before diving into the particulars of which microbes appear in which parts of the body, we will briefly review some of the theory and evidence for how human-associated microbial communities assemble in the first place. As we proceed through the chapter, we will revisit these processes as they apply to different body sites.

The forces that govern microbial assembly appear to be complex and dynamic, involving host genotype, inter-microbial interactions, and environmental factors like diet. As such, many organizing principles for describing and understanding these community assembly processes have been borrowed from macroecology, a field explicitly focused on the complex interactions within and between species, and between species and their environments. Canonically, community assembly can be broken down into four fundamental processes, which then give rise to more complex phenomena. These processes are dispersal, selection, diversification, and drift (Fig. 1). We will describe each of these briefly below (for more thorough review, see Vellend or Costello).[1,2]

2.2 Dispersal

Dispersal refers to the immigration and emigration of microbes from one local habitat to another (Fig. 1A).[2] In the context of the human body, each individual can be thought of as an "island" or "patch" that draws its particular consortium of microbes from the broader pool of microbes available on nearby hosts and environments, or the "meta-community." Thus, the

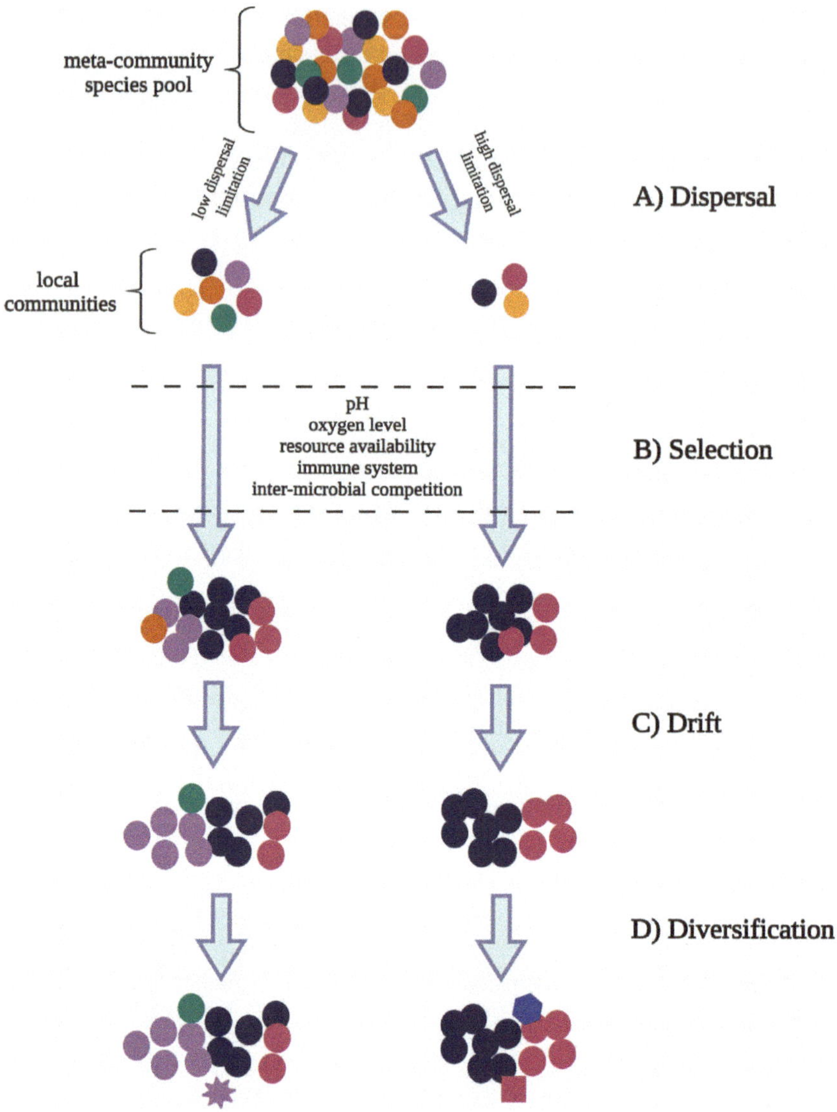

Fig. 1 Summary of ecological processes that impact community assembly. (A) Dispersal refers to the immigration and emigration of microbes between local habitats. The meta-community species pool represents the collection of microbial species present across all local habitats. As species disperse between these habitats, they can be limited by factors such as motility and distance from the source community. (B) Selection refers to the process whereby species better adapted to their environment tend to survive better and produce more offspring. Some potential habitat filters relevant to microbial life are listed between the dashed lines. (C) Drift refers to random fluctuations in species abundances due to stochastic changes in birth and death rates over time. This can disproportionately impact low-abundance species, which can become extinct after a random dip in abundance. (D) Diversification refers to the generation of new genetic variants within a population.

possible diversity of microbes that could colonize our bodies is limited by the diversity of microbes in the meta-community, and by the ability of those microbes to physically move to our "patch." For example, initial colonization of the infant gut microbiota is thought to occur largely by dispersal from the available microbes associated with either the mother or the hospital environment[3]. Depending on mode of delivery (vaginal birth or C-section), the neonate may be differentially exposed to maternal or environmental microbes, thereby limiting dispersal from those sources, and shaping the possible range of microbes the infant might acquire.

2.3 Selection

Selection is a deterministic evolutionary force by which better adapted variants within a population produce more offspring over their lifetime than poorly adapted competitors, and over time, displace those poorly adapted variants from the population (Fig. 1B). These selective pressures often take the form of "habitat filters," or environmental conditions (metabolic resource availability, pH, or adhesion sites, for instance) that limit which taxa are able to survive and grow in a given site. Thus, among the pool of microbes that have dispersed to a site and variants that have arisen via local diversification, those that cannot effectively utilize the available resources, and therefore cannot identify a suitable niche, will be driven from the community. Conversely, microbes that arrive in a site and utilize those resources especially effectively can experience positive selection, in which they grow more rapidly than other populations, and bloom to high levels of abundance in the community.

In the context of the microbiome, selective pressures can also be imposed by interactions with other microbes, host cells, and bacteriophage in the community, which can act to modify the available niche space. Even if a bacterial strain could theoretically survive using resources that are available in a given environment, in order for it to persist in that community, it has to be able to outcompete the other microbes that are likewise vying for those limited resources. Host cells and bacteriophage can also respond to the presence of particular microbes and act to suppress or promote certain populations directly or by modifying the environment.[4,5]

2.4 Drift

Ecological drift refers to random fluctuations in species abundances due to stochastic changes in birth and death rates (Fig. 1C). Practically speaking, this means that low-abundance species are more likely to go extinct from a community, especially in the wake of a perturbation, even if they may have been better adapted to that community than other taxa that happened to be present at higher abundance.[6] Drift can play a particularly prominent role in community re-assembly after a strong perturbation like a course of antibiotics.[7,8]

2.5 Diversification

Diversification describes the generation of new genetic variants within a given population (Fig. 1D). This could occur via mutation, recombination, or as is often the case in bacteria, by horizontal gene transfer. Notably, by contrast with larger-scale organisms, diversification in microbes occurs at a much faster rate owing to large population sizes, rapid reproductive rates, and high mutation rates.[9] Recent data has shown that common human-associated bacteria like *Bacteroides fragilis*, for instance, can evolve specific, within-person adaptations over the course of that person's lifetime, and horizontal gene transfer can allow for antibiotic resistant variants to arise rapidly within a community in the wake of an antibiotic perturbation.[10,11] Diversification therefore provides a way for bacteria already present in a community to adapt to changing selective pressures.

2.6 Applications to the human microbiome

Within the human body, many of these forces may be at play. After all, different body sites constitute highly distinct ecological niches. The lung, for instance, is highly aerobic, whereas the GI tract is predominantly anaerobic; the vagina is comparatively acidic, and the skin is exposed to a barrage of variable conditions like humidity, salinity, and temperature. Within each of these sites, microbes also face different pressures from the host immune system, degrees of exposure to foreign microbes, and patterns of resource availability. Given this environmental variability across body sites, it stands to reason that habitat filtering and deterministic selective processes would play an important role in community assembly, and this is largely borne out by the data: evidence from the Human Microbiome Project has shown that human microbiota are more similar across individuals than across body sites.[12] That is, my gut microbiome is more similar to your gut than my gut is to my elbow, suggesting that selective processes unique to the gut (or elbow) act to shape these communities in predictable ways.

Especially from a biomedical and translational perspective, it is tempting to assume that deterministic forces like selection are the primary drivers of microbiome composition. However, microbial ecologists have increasingly been making efforts to quantify and account for the extent to which more random, stochastic processes like dispersal and drift play into community assembly. For example, Sprockett et al. argue that microbiome researchers seeking to understand community assembly must account for priority effects and historical contingency, in which a community can be shifted toward an alternative stable state by chance early life exposures to a limited subset of microbes, which then occupy and/or modify the available niche space, impacting the ability of late-arriving microbes to colonize.[13] This highlights the possibility that feedbacks between stochastic and deterministic processes can amplify differences in community assembly, allowing communities under similar selective pressures to undergo divergent compositional trajectories. This may explain some of the substantial variation in microbial composition that remains within body sites among healthy individuals.[12]

As we review the microbiome composition and function of the body sites below, we will attempt to highlight what is known about community assembly processes in each of those sites. Filling in the gaps in our understanding of these phenomena is sure to be a central focus of future research efforts.

3. Gastrointestinal tract

3.1 Overview

While all of the body sites reviewed in this chapter are topologically external, the gastrointestinal tract (GIT) is exceptional for the sheer volume of challenges it experiences from the outside world, and for the wide diversity of functional roles it must play as it incorporates defense into the processes of digestion and absorption. Every time we eat, our GIT must identify and neutralize any potentially hazardous elements that hitchhike into the gut with the foreign matter we have ingested, while simultaneously breaking down and converting nutrients from that same matter into a form that can be easily absorbed, all the while sensing and providing feedback to the rest of the body. Thus, the GIT is home to not only a vast array of anatomical structures and biochemical processes that partition the work of digestion and absorption, but also sensory, neurological, and endocrinological networks, as well as complex immunological and ecological systems that are integral to defense and homeostasis.

Given the tremendous diversity of functions that must be performed by the GIT, it is perhaps unsurprising that the gut environment does not consist of a single uniform ecological niche, but rather a heterogeneous collection of distinct habitats along the rostral-caudal axis, which are home to the most abundant and diverse microbiota in the human body.[12] Along this axis, there are several critical gradients: microbial load and diversity increases from the rostral to caudal end, oxygen levels decrease, mucus thickness varies, and pH dips in the stomach and increases distally from there (Fig. 2).[14–16] Each

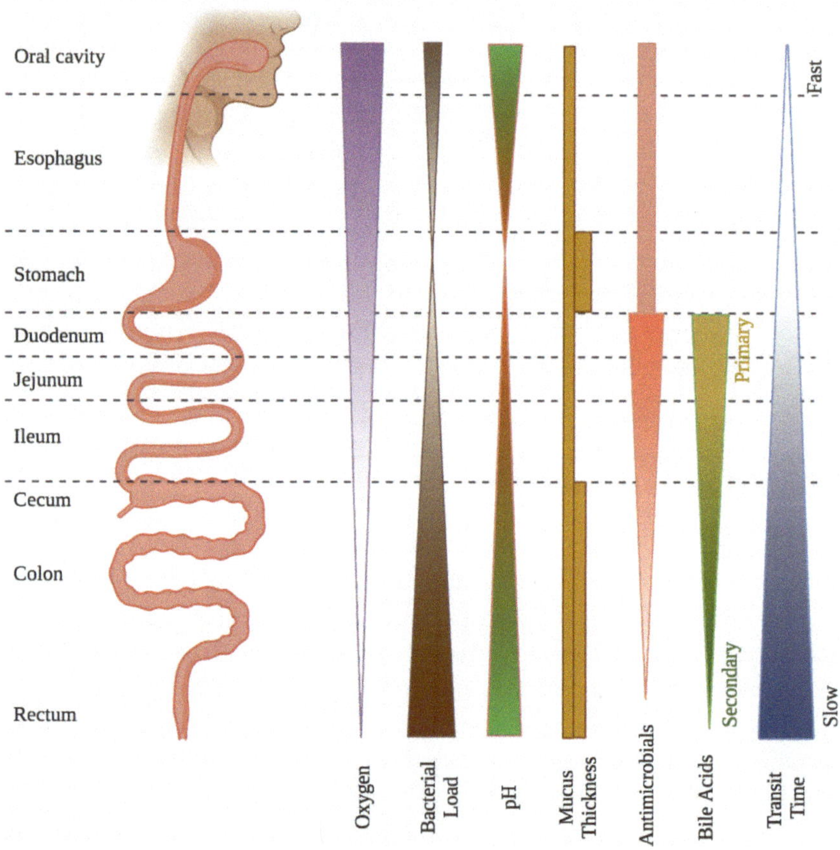

Fig. 2 Ecologically relevant spatial gradients across the gastrointestinal tract. Rostrally to caudally throughout the gastrointestinal tract, a variety of physiological and biochemical features vary in accordance with the distinct digestive, absorptive, defensive, and endocrinological functional roles of each region. These factors, including oxygen level, pH, mucus thickness, anti-microbial production, bile acid production, and transit time, in turn impact the structure and function of regional microbial community membership.

region of the GIT—which we will break down into oral cavity, esophagus, stomach, small intestine, and colon—performs locally specialized digestive, metabolic, immune, and endocrinological functions. This means there is local variation in nutrient bioavailability, GI transit time, and secretion of immune factors and bile acids, among other things.[15,17] These distinctions impact local microbial composition in significant and often predictable ways, and the functions performed by such locally distinct microbial assemblages can then feed back into local host processes for a mutually beneficial symbiosis, or in disease states, a "dysbiotic" microbiota can contribute to impairment of host functionality.

Despite extensive efforts to characterize the microbiomes of healthy individuals, a concrete definition of the "healthy" gut microbiome has been elusive. In part, this is because of substantial microbial diversity even within the same body site among healthy individuals, especially at higher taxonomic resolution (that is, genus, species or strain level).[12,18] Some of that variation may be attributable to idiosyncratic combinations of stochastic and deterministic assembly forces at play. Nevertheless, common definitions of the healthy gut microbiome have come to center around functional potential, rather than taxonomic composition, as well as stability over time, and resilience in the face of perturbation.[19] Here, we describe some such compositional and functional characteristics of the "healthy" regional gastrointestinal microbiota.

3.2 Oral cavity

Serving as the entry point for both the GIT and the respiratory tract, the oral cavity is responsible for a wide variety of functional roles, serving as a critical respiratory conduit, as well as the initiating site for many digestive processes. These include mastication, moistening, the first stages of enzymatic breakdown, as well as taste and other sensory reception, which can play an important role in defense against toxic or pathogenic agents. As the first opportunity for encounters between invading pathogens and the GIT, the oral cavity also plays a pivotal immunological role. Reflecting this diversity of functional roles, the oral cavity is home to a suite of anatomical and functional adaptations that create a rich assortment of spatially distinct habitats, the relative accessibility of which has paved the way for high-resolution spatial sampling of the oral microbiota. Recent work has estimated that some 200–500 unique bacterial species inhabit the mouth, reaching abundances of ~20 million individual cells.[20]

Anatomically, the mouth contains a variety of distinct habitats and topological features, including lips, teeth, gingiva, tongue, cheeks, hard and soft palate, and tonsils. The first key distinction between these features is whether they have shedding or non-shedding surfaces: the only non-shedding surface in the mouth is the dental enamel, while the remainder of the oral mucosa are covered in stratified squamous epithelial cells that regularly turnover and slough off. The non-shedding dental enamel, as a more permanent structure in the mouth, allows for a buildup of plaque, or bacterial biofilms. These multi-species structures are fascinating examples of microbial cooperation and successive community assembly: a base of filamentous *Corynebacterium* tendrils provides a scaffold upon which organized consortia of taxa including *Streptococcus*, *Haemophilus*/*Aggregatibacter*, *Porphyromonas*, *Capnocytophaga*, and *Fusobacteria* attach in "hedgehog" structures according to micron-scale biochemical gradients.[21] Plaques above and below the gum line—that is, supra- and sub-gingival plaques, respectively—also contain distinct microbiota, in part due to the capture and accumulation of nutrients and immune molecules in the crevices between gingiva and teeth.[22]

Although the dominant members of supragingival plaques remain abundant across the mouth, the less abundant members of these communities vary by tooth class (incisor versus molar) and tooth aspect (buccal or lingual).[23–25] Data from Proctor et al. suggests that this is because salivary flow is a major selective force in the oral cavity: as saliva flows from the back to the front of the mouth from various major and minor salivary glands, it creates biophysical gradients of moisture and pH, while also acting to physically shuttle bacteria around, release or clear metabolites from partially digested food, and stimulate secretion of mucin.[26,27] These gradients select for an enrichment of taxa like *Veillonella* on the molars, and *Streptococcus* and *Actinomyces* on the incisors.[25] Microbially-mediated diseases like periodontitis and dental caries are commonly associated with hyposalivation, suggesting that in the absence of salivary flow and associated microenvironmental gradients in the oral cavity, dysbiotic communities can assemble.[26,28,29]

By contrast with the non-shedding dental enamel, the soft tissues of the oral mucosa feature shedding epithelial cells that turnover approximately every 2–3 weeks.[30] Microbial communities on keratinized (hard palate, tongue dorsum, keratinized gingiva) and non-keratinized (buccal and labial mucosa, tongue ventrum, soft palate) tissues are hypothesized to vary according to differences in adhesion mechanisms and proximity to salivary glands.[26,31,32]

As the initial site of ingestion, the oral cavity is also exposed to a number of pathogens, and a major selective force in the assembly of this community is constant basal activation of the immune system.[33,34] Thus, given the diverse assortment of microbes that can disperse into the oral cavity, regular surveillance and basal immune activity, via the IL-17 pathway for instance, is responsible for filtering out potentially pathogenic agents.[35] Symbiotic residents of the oral microbiome such as *Streptococcus*, *Veillonella*, and *Granulicatella* have been shown to contribute to these defensive processes, stimulating an increase in the production of immune effectors like antimicrobial peptides (AMPs) and pro-inflammatory cytokines, and increasing epithelial barrier function and mucosal thickness.[36]

3.3 Esophagus

By contrast with the oral cavity, the esophagus is relatively uniform in anatomy and function. Flanked by the upper and lower esophageal sphincters, the esophagus is primarily responsible for shuttling food from the oral cavity into the stomach. It is covered in non-keratinized stratified squamous epithelial cells and coated in mucus secreted by submucosal glands.

Although a multitude of microbes pass through the esophagus with ingested food in transit to the stomach, these are thought to be largely transient.[37] The more stable resident population of the esophagus resides attached to the mucosal surfaces, which can only be sampled via an invasive upper endoscopy procedure with brushings or biopsy, severely limiting characterization of this body site. Nevertheless, a small number of studies have found that the esophageal mucosa are primarily colonized by ingested microbes from the oral cavity, exhibiting a high degree of compositional overlap, with representation from the same major taxa including *Streptococcus*, *Prevotella*, and *Veillonella*.[37,38] Unlike the oral cavity, the esophagus shows a notable absence of *Spirochaetes*. These patterns are consistent with dispersal-limited community assembly, in which the microbes that colonize the esophagus are a subset of the meta-community present in the oral cavity.

3.4 Stomach

The stomach plays critical digestive, defensive, and endocrinological roles in the GIT. Ingested material that reaches the stomach is mechanically broken down through peristaltic churning, chemically degraded by hydrochloric acid, and enzymatically broken down by zymogen proteases like pepsinogen, which is activated in the acidic stomach environment. The stomach also senses and regulates food intake, motility, and appetite through various mechanisms including stretch mechanoreceptors and production of hormones like gastrin and ghrelin. These functions both impact and are impacted by the gastric microbiota.

These are harsh conditions for microbial life. The low pH of the stomach (~1.5–2.5) constitutes a fairly extreme habitat filter which, in addition to serving as a defense mechanism against ingested pathogens, significantly limits the diversity of commensal microbes that are able to survive there. Historically, this led scientists to believe that the stomach was completely sterile, until *Helicobacter pylori*, an infectious, acidophilic bacterium that contributes to the development of peptic ulcers, was discovered thriving in the gastric mucosa of affected individuals.[39] Recently, however, culture-independent techniques have demonstrated that although overall bacterial load is relatively low in the stomach (~10^1–10^3 CFU/mL), there is representation of diverse phyla, including *Streptococcus*, *Prevotella*, *Veillonella*, and *Rothia*.[40–42] More than 65% of the taxa identified in the stomach are members of the oral microbiota, although many of these, identified in the gastric juice, are thought to be largely transient.[43]

Community assembly in the stomach can be understood as a balance between dispersal from the oral cavity and the duodenum, and selection driven largely by pH, a dominant habitat filter. This is supported by evidence showing that higher pH in the stomach corresponds to overgrowth of taxa likely derived from oral and intestinal source communities.[44] Acidity levels can be impacted by use of medications such as proton pump inhibitors, or even resident microbes: *H. pylori*, which has been documented at ~50% prevalence in asymptomatic adults in the USA, is well known to reduce stomach acidity.[45,46] By breaking down urea into ammonia, a basic substance that neutralizes stomach acid, *H. pylori* can burrow into the gastric mucosa, where it stimulates an immune response that results in atrophy of the cells lining the stomach, further reducing the secretion of acid.[43,47] As the stomach de-acidifies, it becomes more amenable to microbes that disperse from either the oral cavity or the duodenum, ultimately contributing to a "dysbiotic" disease state.[44]

Although *H. pylori* is commonly regarded as an endemic pathogen of sorts in the modern age, it has been hypothesized that, before industrialization, the pro-inflammatory IL-17 response provoked by this highly prevalent microbe served a protective role against other infectious pathogens.[48,49] Since then, germ theory was born, hygiene practices have evolved, and antibiotics have been developed, causing the prevalence of such infectious agents to decrease, and rendering the pro-inflammatory impacts of

H. pylori mismatched to our "clean" environment. Thus, *H. pylori* is now most commonly associated with peptic ulcer disease and gastric cancers (see Blaser for a fascinating review of *H. pylori*'s disparate roles in upper and lower gastric disease).[50]

3.5 Small intestine

The small intestine is the primary site of nutrient absorption in the GIT. It features a single layer of absorptive columnar intestinal epithelial cells (IECs) interspersed with a variety of specialized cells including goblet cells, Paneth cells, microfold cells, and enteroendocrine cells that aid in digestion, defense, and secretion. The apical surface of these IECs maximizes surface area for absorption, featuring crypts, villi, and microvilli, and is coated in a single, loosely attached layer of mucus that is largely free of microbes.[15,51] Compared to the large intestine, the small intestine is more aerobic, more acidic, secretes higher levels of anti-microbial molecules, and has much shorter transit time.[15] Microbes that colonize these regions must not only survive these habitat filters, but must compete for nutrients with both other bacteria and host cells. Therefore, very generally, the small intestine harbors an abundance of fast-growing, facultatively anaerobic bacteria that can utilize carbohydrate energy sources, have robust AMP tolerance mechanisms, and can adhere to mucus, dietary substrate, or epithelial tissues despite rapid GI motility.[15,52]

To facilitate the absorption of nutrients into the IECs, the mucus coating of the small intestine is much thinner than in other regions of the GIT. Thus, the small intestine faces a trade-off between the defensive barrier functions that a thick, secondary mucus layer could provide, and the ability to effectively digest and uptake nutrients from the intestinal lumen. To combat this, the small intestine features special immunological adaptations like Paneth cells in the base of the crypts, which secrete copious amounts of AMPs like lysozyme and α-defensins into the mucus layer, and abundant plasma cells that release secretory IgA into the mucus, thereby inhibiting microbial colonization.[53–55] The small intestinal epithelia overlay specialized gut-associated lymphoid tissue (GALT) and Peyer's patches in the mucosa and submucosa that function in presentation of antigens and activation of the adaptive immune system.[56] As a consequence of this extensive immune activity, some commensal microbes in the gut have been shown to possess unique antigen modifications to escape innate immune mechanisms.[57]

The study of the unique functional roles of the small bowel microbiota in healthy host physiology is an often-overlooked but nascent sub-field within medical microbiome research. While this is largely due to the relative inaccessibility of small intestinal sites compared to fecal sampling, recent work in animal models has begun to establish an essential role for the small bowel microbiota in host metabolism and nutrient absorption. For instance, evidence from Martinez-Guryn et al. shows that, like the well-characterized colonic microbiota, the small bowel microbiota of mice is dramatically impacted by Western-style high-fat (HF) diet, and in fact plays an indispensable role in triglyceride digestion and absorption.[58] This work showed that small intestinal microbes are necessary for the release of pancreatic lipase in response to dietary lipid, indirectly aiding in host lipid digestion. Moreover, the authors showed that metabolites released by taxa like *Clostridiaceae* that bloom under HF conditions interact with absorptive enterocytes of the duodenum and jejunum to enhance the uptake of free fatty acids. Thus, as the host consumes more fat, the microbiome shifts and enables the host to absorb that fat. This is a striking example of the bidirectional crosstalk between host and microbe that is essential for the maintenance of healthy host physiology, especially in the face of rapidly changing conditions like variable diet.

Proximally to distally from the stomach, the small intestine can be divided into three segments, each of which features distinct functional roles, anatomical features, and microbial inhabitants: the duodenum, the jejunum, and the ileum. We will describe each of these regions below.

3.5.1 Duodenum

The duodenum comprises the first 20–30 cm of the small intestine. It is a site of significant secretory and digestive activity, with mucus from goblet cells protecting the intestinal epithelium from the acidic gastric chyme, bicarbonate from the pancreas bringing the pH back up into the neutral zone, and bile and pancreatic enzymes streaming in to aid in the breakdown and absorption of carbohydrates, fats, and proteins. Anatomically, the duodenal epithelia feature crypts and long villi with a microvilli "brush border" packed full of digestive enzymes for secretion and nutrient transporters for uptake. Immunologically, the duodenum features submucosal GALT, although there are fewer Peyer's patches than in the jejunum and ileum.[56]

These features impact the duodenal microbiome in particular ways. First, a fairly high level of swallowed oxygen, as well as oxygen diffused through duodenal tissues, prevents the establishment of more strictly anaerobic microbes. Second, rapid motility of the duodenum may prevent many microbes that disperse from upper segments of the GI tract from having time to adhere or to establish. Thus, bacterial load is fairly low, in the range of

10^1–10^3 CFU/mL, and tends to be dominated by bacteria from the phyla *Firmicutes*, *Actinobacteria*, and *Proteobacteria*, which contain many facultative anaerobes.[59,60]

Primary bile acids (BAs), which are synthesized from cholesterol in the liver, are deposited via the bile duct directly into the duodenum. These amphipathic molecules act as a detergent that aids in the absorption of lipids by emulsifying fat globules. BAs, however, have been shown to exhibit strongly bactericidal activity, and therefore act as a selective pressure for more bile-tolerant microbes throughout the small intestine.[15,61] Bile salt hydrolases (BSH) are enzymes that deconjugate primary BAs, and they are widespread across the microbial inhabitants of the gut.[62] It is hypothesized that these BSH enzymes confer increased tolerance to primary BAs, which could be especially important in the proximal small intestine where primary BAs are most abundant.[63–65]

Exciting new work on extracellular electron transfer (EET) also suggests that iron, which is particularly abundant in the duodenum and jejunum, may be an important metabolic resource for gut bacteria that can utilize EET to produce energy.[66] Analogous to mammalian intracellular oxidative phosphorylation, these bacteria can use extracellular minerals like iron as an electron acceptor, generating electricity and ATP, and supporting bacterial growth on non-fermentable carbon sources. EET pathway orthologues have recently been identified in the genomes of a wide variety of pathogenic and commensal *Firmicutes* species, suggesting that EET may be a significant competitive adaptation in the resource-limited environment of the proximal small intestine.

3.5.2 Jejunum

After the duodenum and proceeding for the next ~2.5 m before the ileum lies the jejunum, which harbors many of the same functional and anatomical features as the duodenum, but with less oxygen, neutral pH, longer villi, increasing numbers of Peyer's patches, slower transit time, and an altered nutrient profile, dependent on what has already been digested or absorbed in the duodenum. Compared to the duodenum, the jejunum performs fewer digestive functions and is more specialized for absorption, serving as the primary absorption site for carbohydrates, small peptides, and vitamins and minerals like folate, calcium, magnesium, and other trace elements.[67]

Because the jejunum is only accessible by endoscopy in humans, studies investigating its microbial composition in healthy individuals are few and far between. Early work related to small intestinal bacterial overgrowth syndrome revealed that the typical bacterial load of the jejunum in healthy individuals is approximately 10^3–10^5 CFU/mL.[68] Subsequent culture-independent analyses have confirmed this estimate, and revealed that compositionally, the jejunum is dominated by a number of facultatively anaerobic genera including *Streptococcus*, *Prevotella*, *Veillonella*, *Rothia*, and *Fusobacterium*, as well as *Escherichia*, and *Klebsiella*.[69] Although mucosal and luminal jejunal samples have not been simultaneously collected from the same individuals for direct comparison, independent studies of the mucosal and luminal jejunal microbiota have identified representation of similar taxa and abundances.[69–71]

3.5.3 Ileum

The ileum is the final segment of the small intestine, stretching ~3 m between the jejunum and the cecum. Compared to more proximal regions of the GI tract, the ileum continues along the same broader environmental gradients: lower oxygen levels, slower transit time, and shorter, broader villi with fewer digestive enzymes. Digesta that arrive in the ileum have been largely depleted of nutrients that are useable by the host, and its primary absorptive roles are uptake of vitamin B_{12} and reuptake of bile acids.[67] While Peyer's patches are sparse in the ileum, it has an abundance of AMP-secreting Paneth cells in its crypts, and a relatively high number of mucus-secreting goblet cells that create a patchy mucus layer of variable thickness.[56,72]

Here, moving distally, the gut environment begins to approach the optimal conditions for microbial growth that characterize the colon: because the host has already taken up the majority of nutrients it is capable of using, microbes that inhabit the distal ileum are released from competition with host cells for remaining resources. Moreover, the trade-off between absorptive capacity and defense has been largely relieved, as nutrient absorption is less important in the ileum, and therefore immune mechanisms that are thought to reduce microbial density in the proximal small intestine are relaxed.[15] Consequently, we see an increase in microbial load by several orders of magnitude compared to the jejunum, with estimates ranging from 10^3 to 10^8 CFU/mL.[56]

Because the ileum is difficult to sample without invasive procedures, insights into its composition have mostly been gleaned from patient populations with ileostomies, or those undergoing radical cystectomy.[52,73,74] With the caveat that such patient populations typically have underlying health conditions that may or may not impact the composition of the ileal microbiota, these studies have revealed that the ileum is home to both facultative and obligate anaerobes, including taxa from the genera *Streptococcus, Granulicatella, Actinomyces, Solobacterium, Rothia, Gemella,* and *TM7(G-1)*.[52,73] This is significantly different than the colonic microbiota, with less representation of strictly anaerobic families like *Clostridiaceae*, *Peptostreptococcaceae*, or *Eubacteriaceae*, and an enrichment for facultative anaerobes like *Streptococcaceae* and *Lactobacillaceae*, although this may be due in part to differences in oxygen tension in ileostomy patients.[73]

Although immune release is thought to be one reason why bacterial load in the ileum is higher than in other parts of the small intestine, mucosal immunity still plays a pivotal role in shaping microbial composition in the ileum, and vice versa.[75] Abundant production of AMPs such as cathelicidins, C-type lectins, and defensins by Paneth cells create a defensive border between epithelial cells and microbial inhabitants. However, production and secretion of these immune effector molecules is dependent upon stimulation by commensal microbes: germ-free mice show reduced levels of AMPs like the Gram-positive-targeting REGIII-γ, which is restored upon conventionalization with cecal contents from specific-pathogen free mice.[76] Immune stimulation by commensal microbes can play an important role in colonization resistance against certain pathogenic organisms, as well. For instance, mice lacking segmented filamentous bacterium, a microbial resident unique to the terminal ileum, fail to develop Th17 cells, and are consequently highly susceptible to infection by pathogens like *Citrobacter rodentium* and *Salmonella typhimurium*.[77,78]

Ileal and colonic microbes also play an important role in the maintenance of host circadian rhythmicity. Circadian rhythms refer to periodic 24-h oscillations in gene expression across body tissues, which act to coordinate homeostatic functions including the sleep-wake cycle, feeding behavior, and glucose and lipid metabolism.[79] By temporal partitioning of various catabolic and anabolic processes, circadian rhythms are able to curtail inefficiency in resource use by the body, and dysregulation of these rhythms has been associated with chronic metabolic disorders like diabetes and obesity.[80] Although the central circadian clock is located in the suprachiasmatic nucleus of the brain and is primarily entrained by photic cues, oscillatory dynamics in peripheral tissues like the gut can also be entrained by feeding behaviors and microbial stimuli.[81,82] In particular, Wang et al. have shown that, in the small intestine, rhythmic, circadian expression of *Nfil3* is regulated by the intestinal microbiota, and that antibiotic-induced loss of *Nfil3* rhythmicity is associated with increased body weight and fat storage.[83]

3.6 Colon

The colon, or large intestine, begins at the cecum, a small pouch just distal to the ileum, and extends approximately 150 cm to the end of the GI tract at the rectum and anus. This region is home to the most abundant assortment of microbes in the human body, containing orders of magnitude more microbes than all other body sites combined with bacterial density on the scale of 10^{10}–10^{12} CFU/mL.[59] This is promoted by the specific anatomical and functional features unique to this region.

Like the small intestine, the colon features a single layer of polarized columnar intestinal epithelial cells with divot-like crypts, but unlike the small intestine, the colonic epithelium lacks villi and microvilli. Goblet cells in the colonic epithelia secrete a thick, two-layered mucus coat, with a dense inner layer that stays largely sterile, and a loose outer layer that harbors an abundance of specialized microbes.[16,51] The colon is highly anaerobic, and the digesta that pass through consist primarily of complex polysaccharides and fibers that could not be digested by host processes, as well as trace nutrients and any remaining bile acids that were not absorbed in the ileum. Motility in the colon is much slower, with typical transit time of up to 30 h.[84]

These features allow microbes to thrive for a number of reasons. First, fermentative metabolism is widespread in bacteria, and the anaerobic environment and abundance of fibrous substrates in the large intestine create optimal conditions for this process. Second, the slow transit time through the colon allows microbes plenty of time to adhere, consume, multiply, and expand in physical space. Thus, because they are not being regularly flushed through the system, microbes in the colon can accumulate to higher levels. Third, the colon provides a number of spatially distinct niches that can support different microbial communities, from the colonic crypts, to the outer mucus layer, to the inter-fold regions of the lumen and central lumen. This spatial heterogeneity allows for niche partitioning of limited resources, increasing the total level of diversity that can be supported.[85] Finally, as long-term, co-evolved symbionts, microbes in the colon have developed unique adaptive mechanisms of immune evasion and modulation to avoid being actively removed from the gut by host defense mechanisms.[57,86]

Microbial synthesis of short-chain fatty acids (SCFAs) through fermentative metabolism of complex polysaccharides is the canonical mutualistic function of the gut microbiome: this was perhaps the first major physiological function discovered to depend wholly on the gut microbial "organ."

In this catabolic process, fibrous substrates like cellulose, pectin, inulin, and high-amylose starch, or host-derived mucosal glycans, are fermented anaerobically by microbial symbionts across the *Firmicutes* and *Bacteroides* phyla, releasing SCFAs like acetate, proprionate, and butyrate.[87] These SCFAs are salvaged by host tissues, contributing an estimated 5–15% of the total caloric requirement for humans.[88] While all of these SCFAs can feed into various host and microbial metabolic pathways, butyrate in particular serves as the preferred energy source for colonocytes.[89,90] These cells oxidize butyrate into CO_2, thereby depleting the oxygen supply in the colon and promoting an anaerobic state that is important for pathogen resistance, immune homeostasis, and of course, the growth of butyrate-producing anaerobic microbial populations, in a classic positive feedback loop.[87,90–92]

Microbial composition, and therefore fermentation and metabolism, in the colon can be dramatically affected by the components of the diet consumed by the host. One early study examined human subjects who switched between an animal- and plant-based diet for a designated period of time, demonstrating that microbial composition changes rapidly and reversibly in response to dietary components.[93] Additional work has extensively documented the profound impact of high-fat and low-fat diet on fecal microbial communities, with high-fat diet shifting communities toward a greater ratio of *Firmicutes:Bacteroidetes* taxa.[94] At somewhat higher nutrient-resolution, recent work has shown that administration of specific complex polysaccharides can promote the growth of specific *Bacteroides* species.[95,96]

Whereas resource availability in the colonic lumen varies depending on dietary intake, many colonic microbes reside in and consume host-derived components of the mucosa, which may remain somewhat more stable across dietary behavioral patterns.[15] The glycoprotein that makes up the majority of the colonic mucus, Mucin 2 (MUC2), is coated with a diverse assortment of O-linked glycans, oligosaccharides that can be cleaved from MUC2 and metabolized to support the growth of various specialized microbial taxa like *Akkermansia muciniphila.*[97,98] Different microbes vary in their capacity to penetrate and adhere to the mucus layer, as well as in their tolerance to AMPs and oxygen that diffuse outward through the mucus from the underlying epithelial cells.[99,100] Thus, the mucosal niche experiences unique selective pressures compared to the luminal niche in the colon, and consequently supports an enrichment of aerotolerant, asaccharolytic protein-metabolizing species from the phyla *Actinobacteria* and *Proteobacteria* compared to the lumen.[100,101] Within the mucosa, as well, researchers have documented further spatial niche partitioning. For instance, *Acinetobacter* species are particularly effective at navigating through the mucus layer and its associated biochemical gradients to bind directly to the IECs of the colonic crypts.[102]

Like the mucosa, the lumen itself also harbors significant spatial heterogeneity, much of which has been overlooked by the use of fecal sampling in the majority of colonic microbiome studies. Evidence in mice demonstrates that the characteristic mucosal folds of the colonic walls create distinct "inter-fold" regions in the intestinal lumen, which are enriched for taxa in the families *Lachnospiraceae* and *Ruminococcus*. These taxa are thought to benefit both from the local accumulation of mucus from the epithelia and from an environment that is relatively protected from the flow of other luminal contents.[17,103] The digesta of the central lumen, by contrast, are dominated by strictly anaerobic, saccharolytic taxa from the families *Bacteroidaceae, Enterococcaceae, Prevotellaceae,* and *Rikenellaceae.*[103]

Proximally to distally, the microbial inhabitants of the colon exhibit locally specialized functions that seem to broadly reflect resource availability. Because digesta entering the colon have higher concentrations of complex polysaccharides and bile acids, fermentation and bile acid metabolism are largely localized to the cecum and proximal colon.[65,75,104]

Bile acid metabolism extensively impacts microbial composition in the colon. Primary BAs not absorbed in the ileum enter the cecum and colon, where they can be deconjugated via BSH enzymes that are widespread among commensal microbes, and/or converted into secondary BAs by select bacterial species in the *Clostridia* class with 7-α-dehydroxylase enzymes.[62,105] BAs are bactericidal to many taxa, and therefore the processes of deconjugation and conversion can serve as a defensive measure for microbial community members.[61,106] Secondary BAs are typically harmless to the bacteria that synthesized them, but can still exert toxic effects on other microbial taxa. Thus, under conditions with high concentrations of primary BAs, as in the proximal colon, we would expect that the ability to convert primary BAs into secondary BAs would confer a species-specific growth advantage. This prediction finds support in the literature. One study in rats showed that the size of the BA pool can exert an extremely strong selective pressure on the colonic microbial community: rodents that received a dietary BA supplement exhibited dramatic, phylum-level shifts in microbial composition, particularly favoring *Clostridia* species that can convert primary BAs into secondary BAs.[107] As high-fat diet induces the production of more bile salts for lipid absorption, these bile salts in turn impact the colonic microbiota.[93,108]

The composition and size of the BA pool, which both affect and are affected by microbial metabolism throughout the colon, likewise have profound impacts on host health and metabolism. For instance, deconjugated bile acids produced by the actions of microbial BSH enzymes can disrupt micelle formation and inhibit absorption of cholesterol and other lipids across intestinal membranes, with implications for cardiovascular disease.[65] Secondary BAs generated by commensal microbes can activate nuclear receptors like Farnesoid X Receptor (FXR) and G-protein coupled receptors like TGR5 to communicate across host tissues and regulate primary bile acid synthesis and degradation, glucose and lipid metabolism, and energy homeostasis.[109–111] Finally, microbially-synthesized bile acids are increasingly being investigated for their roles as mediators of inflammatory disease.[108,112]

The inflammatory state of the host, in and of itself, plays a large role in the composition and function of the colonic microbiome, reflecting the delicate homeostatic balance between the immune system and the microbiome in this densely populated region of the GIT. Microbes that colonize the healthy gut have evolved a diverse array of mechanisms to avoid triggering an immune response. For instance, many commensal strains, particularly from the *Bacteroides* genus, have evolved a modification to the lipopolysaccharide (LPS) on their outer membrane that renders it invisible to host-derived AMPs.[57] Other commensal taxa have evolved more explicit immunomodulatory strategies. A component of the polysaccharide capsule of *Bacteroides fragilis*, for example, stimulates regulatory T cells to produce immunosuppressive interleukin-10, allowing *B. fragilis* to colonize the mucosal niche.[86,113] Although it is unclear whether a dysbiotic microbiome triggers a pathologic immune response or vice versa, many studies have documented a compositionally distinct microbiome under inflammatory conditions that is broadly characterized by reduced representation of the *Firmicutes* and *Bacteroidetes* phyla, and increased abundance of *Actinobacteria* and *Proteobacteria*.[114–117]

4. Skin

4.1 Overview

As an enormous organ that is fully external, the skin and its associated microbiome may be exposed to the most variable conditions of any body site. Microbes can arrive on the skin from any source community that the host encounters, experiencing very low dispersal limitation, and although the skin exhibits distinct biogeography across both regional and local scales, the conditions in these sites often vary in accordance with environmental conditions and host behaviors. Generally, the skin is dry, cool, aerobic, and acidic, although across regional structures and folds, there is variation in these physicochemical parameters, as well as in the density of micro-scale topological structures. Certain regions, like the axilla, for instance, tend to be warmer and more moist, with densely packed sweat glands and hair follicles. Even this pocket, however, is subject to temporal and behavioral variability—an individual that is exercising on a hot day will have a different axillary environment than an individual that is sitting around on a cold day. And while micro-scale topological features like hair follicles or sweat glands tend to act as more consistent habitat filters, even these sites are subject to frequent perturbations due to variable host cosmetic and hygienic behaviors.

The skin, therefore, emerges as a site with (1) very low dispersal limitation, and an extensive meta-community species pool, (2) numerous regional and local habitat filters for niche differentiation, and (3) high rates of perturbation. These criteria are among the most widely cited in the ecological literature as mechanisms for maintenance of diversity.[85,118,119] Such characteristics pave the way for stochastic processes like dispersal and drift to interact with the diverse selective pressures of the skin, creating a wide variety of possible community outcomes across individuals. And indeed, data from the Human Microbiome Project has shown that, although the GIT exhibits comparable alpha diversity, or within-site diversity, the skin has the highest beta diversity of all body sites, which means that it is the most variable from person to person.[12,120] Furthermore, longitudinal tracking of the skin microbiome reveals that it is also the least temporally stable body site, with substantial intra-individual variation over time.[120,121]

4.2 Skin anatomy, physiology, and microbiome composition

Structurally, the skin consists of shedding layers of stratified, keratinized squamous epithelial cells with interspersed structures like hair follicles, pores, and sebaceous glands. An estimated 10^{11} microbial cells blanket the surface of the epidermis, descending into the pores, glands, and follicles.[122,123] These structures are not only important sources of physical spatial heterogeneity, but their physiological functions contribute to local patchiness in moisture, acidity, and salinity across the skin, creating distinct, and often selective, ecological niches (Fig. 3).[124]

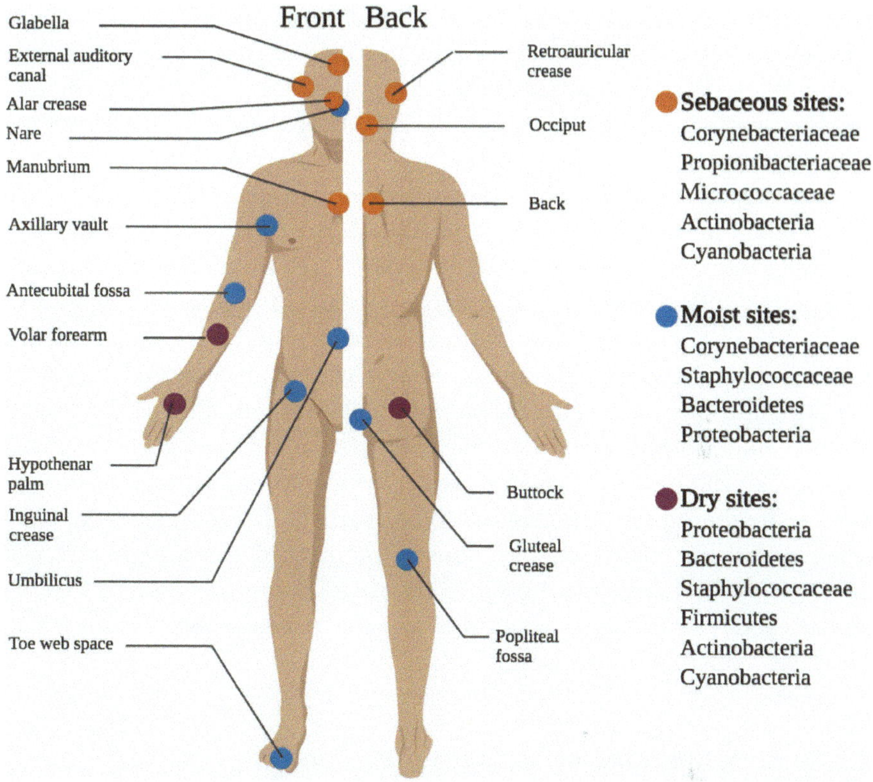

Fig. 3 The skin microbiota varies by region. The composition of the skin microbiome varies in accordance with regional and local topographical features of the skin. Broadly, these sites can be categorized by whether they are moist, dry, or sebaceous, and each of these categories features a distinct subset of microbial taxa that are uniquely well-adapted to those conditions. *Adapted from Grice et al. Grice EA, Segre JA. The skin microbiome.* Nat Rev Microbiol. *2011;9(4):244–253. doi:10.1038/nrmicro2537.*

Apocrine sweat glands, for instance, recruit *Corynebacteria* species. In response to adrenaline, these glands release odorless steroids, acids, and other volatile secretions, which are consumed and metabolized by *Corynebacteria* into the characteristic malodorous compounds of sweat.[125–127] Sebaceous glands, which are connected to hair follicles and secrete a lipid-rich substance called sebum that protects and lubricates the skin, are also highly selective for particular bacteria. Skin regions with dense sebaceous glands, like the face, chest, and back, show the lowest beta diversity compared to other skin sites, consistently harboring the same bacterial taxa across individuals.[128] These glands recruit lipophilic species like *Propionibacterium acnes*, which consumes and degrades the lipids that make up sebum.[129,130] This releases free fatty acids as a byproduct, contributing to the baseline acidification of the skin, a primary barrier against pathogens.[131,132] Sebaceous sites are also home to fungal commensals like *Malassezia*, another lipophilic microorganism.[133]

The density of such topological structures, as well as the broader-scale anatomy of different skin regions, impacts the physiology and environment in those locations. Temperature and humidity, for example, vary dramatically across different anatomical structures. Less exposed, folded regions of the skin like the axillary vault, inguinal crease, gluteal crease, or umbilicus, tend to be both warmer and more moist than other more exposed regions of the skin, and they consequently recruit an abundance of humidity-loving bacteria like *Corynebacteria* and *Staphylococci*.[121,128] Drier, more exposed regions like the limbs experience greater temperature fluctuations, as well as more frequent perturbations than other parts of the body, possibly contributing to the relatively low bacterial biomass in those regions.[122,134] Despite this, such exposed regions are among the most diverse sites in the human body, with representation from the *Actinobacteria*, *Proteobacteria*, *Firmicutes*, and *Bacteroidetes* phyla.[121,128,135]

4.3 Functional roles of the skin microbiome

Functionally, the skin microbiome appears to be somewhat less integral to healthy physiology than the microbiomes of other body sites. Most skin symbionts are thought to be commensal, with relatively few examples of obligate mutualism in which the host depends on its microbes to fulfill a particular function. This is perhaps unsurprising in an environment as variable as the skin—the selective pressures are inconsistent, and therefore selection cannot act in a directed manner. Nevertheless, there is emerging evidence that the skin microbiota plays a role in stimulating and educating the host immune system: secretion of the AMPs like cathelicidin and β-defensins, as well as the production of complement, are induced by commensal stimulation of host innate immune receptors.[136–138] Moreover, in mice, recent evidence showed that skin commensals can tune and regulate the adaptive immune response through their ability to shift IL-1 production.[139,140] Finally, commensal microbes like *S. epidermidis* have demonstrated the ability to directly inhibit the growth of pathogenic organisms like *Staphylococcus aureus* and Group A *Streptococcus*.[141]

5. Vagina

5.1 Overview

The vaginal microbiome is among the least diverse human body sites, and has shown one of the tightest, most consistently mutualistic symbioses with the host. While vaginal microbial residents consume host-derived materials from sloughed cells and secretions, the host is thought to benefit from the protective role that vaginal microbes may play against colonization by an assortment of pathogens, including those that might otherwise cause yeast infections, sexually transmitted infections, and urinary tract infections. These tight associations between the host and specific microbial taxa, and mechanistically well-understood mutualisms suggest that the vaginal microbiome may be an exceptional example of host-microbe coevolution.

5.2 Vaginal anatomy, physiology, and microbiome composition

The vaginal epithelium consists of stratified, glycogen-filled, non-keratinized squamous cells that are rapidly shed, with the top layer turning over approximately once every 4 h.[142] It is largely anaerobic, and quite acidic (pH 3.5–4.5) in reproductive-age women. Although vaginal epithelial cells can produce lactic acid, substantial evidence suggests that mutualistic bacteria, predominantly *Lactobacilli*, are the primary source of this acidity, converting glycogen stores from shedding epithelial cells into lactic acid through anaerobic fermentation processes.[143]

This ecosystem, in most healthy reproductive-age women, is dominated by *Lactobacilli*.[144] Historically, the presence of abundant *Lactobacilli* and concomitantly acidic pH were considered signatures of a "healthy" vaginal microbiome, whereas a more diverse microbiome with higher pH has been correlated with preterm birth, "dysbiotic" conditions like bacterial vaginosis (BV), and the presence of vaginal pathogens like *Gardnerella* or *Trichomonas*.[145–148] However, broader culture-independent sampling across different demographic groups has revealed that approximately 20–30% of healthy women have a more diverse, non-*Lactobacillus*-dominated microbiome, once again muddling our understanding of what exactly defines "health" in this ecosystem.[144,149] This work has identified five distinct "community state types" (CSTs) among healthy women, four of which are each dominated by a different species of *Lactobacillus* (*L. crispatus*,

L. gasseri, *L. jensenii*, *L. iners*), and the fifth (CST-IV) by a diverse consortium of facultative and strict anaerobes, including microbes from the genera *Atopobium*, *Corynebacterium*, *Anaerococcus*, *Peptoniphilus*, *Prevotella*, and *Gardnerella*.[144,150] Demographically, black and Hispanic women are more likely to have a CST-IV consortium, whereas white and Asian women are more likely to have any of the other types.[144,149,151]

5.3 Functional roles of the vaginal microbiome

The native vaginal microbiome is thought to contribute significantly to pathogen resistance in the vagina by a number of mechanisms. First, lactic acid produced in copious amounts by resident *Lactobacilli*, and in lesser amounts by host cells and other microbial inhabitants, is inhospitable or directly inhibitory to many invading pathogens including sexually-transmitted pathogens like *Chlamydia trachomatis*, *Neisseria gonorrhoeae*, and HIV.[152–154] Different *Lactobacillus* species produce different amounts and isomers of lactic acid, contributing variably to the lowering of vaginal pH, and therefore their protective capacity.[155] Second, lactic acid has been shown to exhibit protective immunomodulatory properties. Whereas a pro-inflammatory milieu is associated with BV and STI acquisition, work from Hearps et al. have demonstrated that lactic acid can induce production of the anti-inflammatory cytokine IL-1RA, and inhibit production of pro-inflammatory cytokines IL-6 and IL-8.[156] Third, vaginal *Lactobacilli* can produce anti-microbial bacteriocins that specifically target pathogens like *Klebsiella*, *Gardnerella*, *E. coli*, and *Enterococcus faecalis*.[157,158] Finally, it has been hypothesized that native vaginal microbes help to prevent colonization by pathogens in part by competitive exclusion: the native vaginal microbiota is better adapted to the vaginal environment, and can more efficiently extract the limited resources there, preventing potential invaders from growing and establishing.[159]

5.4 Community assembly in the vagina

It is an open question as to why certain healthy women have a *Lactobacillus*-dominated vaginal microbiome and others do not, and among those who do, why different *Lactobacillus* species come to dominance. Unlike the GI tract where it is easy to speculate about the selective pressures and habitat filters that shape community assembly, clear habitat filters have not been identified in the vagina. Many have hypothesized that *Lactobacilli* are selected for their ability to metabolize glycogen-derived resources, but this still fails to explain why other glycogen-metabolizing bacteria aren't more prominent in these communities, or why only four out of over 130 known *Lactobacillus* species dominate the vaginal microbiota of 70–80% of healthy women.[160] Functional genomic efforts to identify habitat-related traits specific to these four *Lactobacilli* have been unsuccessful.[161] One speculative explanation for single-strain dominance could be that priority effects allow a single, early-arriving *Lactobacillus* species to establish, conferring a spatial and numerical growth advantage that limits the ability of late-arriving species to compete, but this possibility requires experimental investigation.

Unfortunately, there exist no established animal models for studying the vaginal microbiome: differences in reproductive strategies that often facilitate research throughput (i.e., more frequent reproduction and greater numbers of offspring) dramatically affect the anatomy and physiology of the reproductive tract and its associated microbiome. Somewhat surprisingly, no other mammals, including closely related mammals like non-human primates, display such high abundances of *Lactobacilli*, or an even passably human-like community composition.[162] Thus, mechanistic understanding of vaginal community assembly may remain elusive until better experimental systems are established.

5.5 Vaginal microbiome stability and resilience

The physiology of the vagina is uniquely time-dependent compared to other body sites, given the cyclic nature of reproductive physiology: hormone levels oscillate with regular menstrual cycling, and pregnancy causes a dramatic series of anatomical and functional changes to unfold in an orderly progression over time. In addition to these longer-scale chronic disturbances to the vaginal microbiome, the vagina is often subject to more acute perturbations due to sexual behaviors, use of contraceptives, lubricants, and menstrual products, and other behavioral practices. Given the regularity of both chronic and acute perturbations to this community, whether or how stable the vaginal microbiome is has been a fundamentally important question. Interestingly, the healthy vagina is both remarkably stable and highly resilient. Acute behavioral disturbances to the vaginal community, as well as regular monthly menstrual cycling, seem to induce only mild shifts in microbial community membership, and most communities ultimately return to their baseline compositional state.[163–165] Similarly, longitudinal studies of the vaginal microbiome throughout pregnancy have revealed that, although there is reduced representation of CST-IV among pregnant women, suggesting a pregnancy-induced shift toward *Lactobacillus* dominance, over the course of pregnancy, the vaginal microbial community remains exceptionally stable compared to non-pregnant women.[149,166,167]

6. Respiratory tract

6.1 Overview

Historically, although the upper respiratory tract (URT) was well known to harbor large numbers of bacteria, it was thought that the lower respiratory tract (LRT) was sterile except when in a state of active disease. With the advent of culture-independent sequencing, this misconception has been revised, and we now know that there is in fact a low biomass but significant healthy LRT microbiome. The characterization of the respiratory tract microbiome and its dynamics over time and across disease states is an emerging field that is slowly bringing nuance to long-held pulmonological dogma.

6.2 Upper respiratory tract anatomy, physiology, and microbiome composition

The URT consists of the nares, nasal and oral cavities, pharynx, and upper larynx. For a description of the microbiota of the oral cavity, see the gastrointestinal tract section above.

Although the microbiota of the pharynx generally reflects the microbiota of the oral cavity, the nares and nasal cavity harbor a distinct subset of microbes. Anatomically, the nares features more skin-like features than the rest of the nasal mucosa, including keratinized, stratified squamous epithelial cells, sweat and sebaceous glands, and follicles through which coarse, specialized hairs called vibrissae emerge. Although air is moistened and warmed as it proceeds through the nasal cavity, air that enters the nasal vestibule is at ambient temperature, and therefore this habitat is cooler and drier than the rest of the nasal cavity. Microbial communities inhabiting the nasal vestibule are highly reflective of skin communities, featuring abundant *Corynebacteria*, *Staphylococcus*, and *Propionibacteria*.[121,168]

Located deeper within the nasal cavity, the nasal mucosa have several significant anatomical and physiological features that distinguish them from both the nasal vestibule and the oral cavity. Compared to the nasal vestibule, the nasal mucosa are warmer and more moist. The epithelium transitions from keratinized stratified squamous cells to pseudostratified, ciliated columnar cells that are blanketed in a flowing layer of mucus. In conjunction with this mucus, the cilia perform an up-and-out sweeping motion throughout the respiratory tract to clear particulate matter. This region is dominated by *Actinobacteria*, including *Corynebacterium* and *Proprionibacterium* as in the anterior nares, but had greater representation of species from *Proteobacteria*.[169,170] This suggests that the nasal cavity may be seeded by microbes that have dispersed from the nasal vestibule, but has distinct habitat filters that select for a unique subset of these microbes.

6.3 Lower respiratory tract anatomy, physiology, and microbiome composition

The LRT comprises the trachea, bronchi, bronchioles, and alveoli. These environments are uniformly quite nutrient-poor and aerobic, although a number of significant anatomical and physiological transitions take place from the proximal to distal end. Most of these tissues, from the nasal cavity through the bronchioles, make up the "conductive" portion of the respiratory tract, responsible for bringing air into the lungs and clearing all other particulate matter, while the alveoli make up the "respiratory" portion, and are responsible for gas exchange. The epithelia of these regions largely reflect these distinct physiological roles.

The trachea and bronchi, like the nasal cavity, feature ciliated pseudostratified columnar epithelia. At the bronchioles, this epithelium transitions into simple columnar cells, which flatten into cuboidal epithelium, and eventually progress to a thin lining of squamous cells in the alveoli. An assortment of secretory cells are interspersed throughout the LRT, releasing mucus which coats the epithelial surface and traps secreted anti-microbial peptides and immunomodulatory molecules.[171] Airway mucus is thickest in the proximal regions of the LRT, thins toward the distal bronchioles, and is eventually replaced by surfactant in the alveoli.[171] Proximally to distally, the LRT also features gradients of oxygen tension, pH, temperature, and density of inhaled particles.[172–174]

Although it seems likely that this spatial heterogeneity could give rise to local differences in microbial composition across the LRT, in humans, current sampling techniques do not have the spatial resolution to investigate this possibility. Samples of the LRT microbiome are typically collected through a process called bronchoalveolar lavage (BAL), in which a bronchoscope is inserted into the bronchioles, and sterile saline is introduced and recollected with whatever microbial community members were washed into it. Microbes in different layers of the epithelial mucus, or those that might appear in the alveolus versus in the bronchiole, are all mixed into a single homogenized solution, hampering detection of distinct communities across microhabitats.

On a somewhat broader spatial scale, however, BAL data gathered from different lobes of the lung seem to suggest that, despite observed environmental gradients across this same regional scale, microbial communities are nearly indistinguishable.[175] Sequencing-based analysis of samples from the lingula, right middle lobe, left and right upper lobes, and supraglottic space of the trachea all featured abundant aerotolerant *Prevotella, Veillonella,* and *Streptococcus* species.[176] Samples from the LRT sites of each individual were more similar to the URT source community from that same individual than the same LRT site in other individuals, likely reflecting the unique, individual-based meta-community from which the LRT community is assembled.[176] These results suggest that habitat filters may not play as important a role in community assembly in the lung as neutral, dispersal-based processes.[177]

6.4 Community assembly processes in the healthy and diseased lung

Given these findings, Dickson et al. have proposed a model of microbial community assembly in the lung adapted from island biogeography theory, a classic ecological model (Fig. 4).[118,176] In this model, Dickson et al. suggest that diversity in the lung is a balance between stochastic immigration from the meta-community species pool (the URT) and extinction events. The closer a site is to the source community, the greater the rate of immigration, and the higher diversity that site will maintain. Under this model, immigration into the LRT occurs when microbes from the URT physically migrate down the oropharynx, are inhaled as aerosolic particulate matter, or are micro-aspirated during sleep. Extinction events occur due to various forces like coughing, mucociliary sweeping, and immune clearance. Since communities at different sites are drawing from the same meta-community pool, there will be significant overlap between sites—each local community should theoretically be a subset of the same source community—but we expect decreasing species richness with distance from the source community. While higher resolution spatial data on the LRT microbiome would be invaluable in testing and refining this hypothesis, Dickson et al. found that species richness did decrease with distance between each sampling site and the URT, supporting this expectation.[176]

Interestingly, as patients transition from health to disease, the relative importance of neutral and selective processes seems to shift: the lungs of

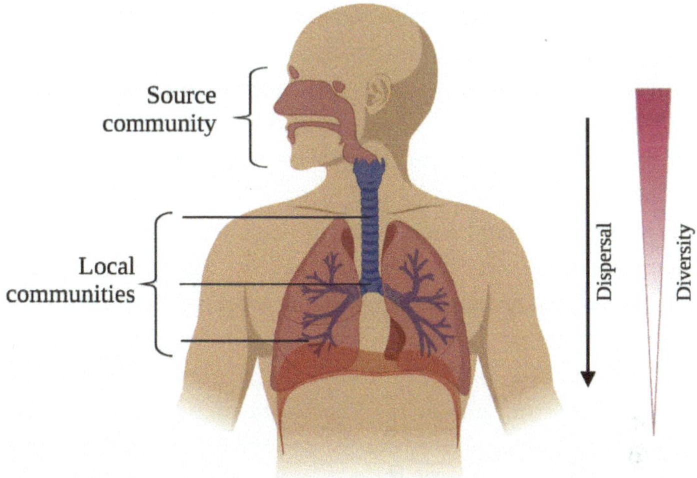

Fig. 4 Adapted island theory of healthy lung microbiome assembly. According to this model proposed by Dickson et al. the upper respiratory tract (URT), and particularly the oral cavity, constitute a source community from which microbes disperse to the lower respiratory tract (LRT). Microbes present in each local habitat of the LRT are therefore a subset of the microbes present in the URT source community. Sites deeper within the LRT face greater dispersal limitation, as they are physically farther from the source community, and therefore diversity decreases with distance from the URT.

patients with advanced lung disease like cystic fibrosis or COPD exhibit distinct spatial heterogeneity in microbial community structure that is absent in healthy individuals.[177–179] This could be due to either the proliferation of native or invading microbes that thrive under disease conditions that are locally exacerbated, or by the increased sensitivity of existing microbes to such local conditions, and their resulting extinction. Generally, injury and inflammation of the respiratory tract lead to increased temperature and increased production of mucus, which in turn creates anaerobic pockets that can support the growth of particular community members or pathogens.[180,181] Locally severe disease may create such conditions in distinct regions, allowing for these new habitat filters to shape community composition there.

7. Conclusions

It is only fitting that a body with as many specialized functions as ours features accordingly diverse, specialized microbiota across those different anatomical and functional sites. In this review, we sought to emphasize that human-associated microbial communities can be shaped by a number of different processes, like dispersal, selection, and drift, and that these processes can contribute in variable measure to the assembly and composition of different body sites. Hopefully, a stronger understanding of these forces will one day allow us to more strategically manipulate these communities toward health.

We also wanted to emphasize that, although much of the data presented here has been instrumentally defined as "healthy" due to the absence of clinical symptoms among study participants, a more precise definition of the "healthy" microbiome is necessary before the microbiome field can transition into a more translational era. Given the degree of microbial diversity even among healthy individuals, an important focus of the field should be the development of positive criteria for salient functions or characteristics of a "healthy" microbiome across body sites, rather than simply the absence of disease. As our definitions, frameworks, and mechanistic understandings of the microbiome continue to develop and refine, our capacity to engineer this complex ecological network for positive health impacts can only expand.

Acknowledgment

MK was supported by MSTP Training Grant (T32GM007281).

References

1. Vellend M. Conceptual synthesis in community ecology. *Q Rev Biol.* 2010;85 (2):183–206. https://doi.org/10.1086/652373.
2. Costello EK, Stagaman K, Dethlefsen L, Bohannan BJM, Relman DA. The application of ecological theory toward an understanding of the human microbiome. *Science.* 2012;336(6086):1255–1262. https://doi.org/10.1126/science.1224203.
3. Dominguez-Bello MG, Costello EK, Contreras M, et al. Delivery mode shapes the acquisition and structure of the initial microbiota across multiple body habitats in newborns. *Proc Natl Acad Sci U S A.* 2010;107(26):11971–11975. https://doi.org/10.1073/pnas.1002601107.
4. Chibani-Chennoufi S, Bruttin A, Dillmann ML, Brüssow H. Phage-host interaction: an ecological perspective. *J Bacteriol.* 2004;186(12):3677–3686. https://doi.org/10.1128/JB.186.12.3677-3686.2004.
5. Sun CL, Relman DA. Microbiota's "little helpers": bacteriophages and antibiotic-associated responses in the gut microbiome. *Genome Biol.* 2013;14(7):127. https://doi.org/10.1186/gb-2013-14-7-127.
6. Gilbert B, Levine JM. Ecological drift and the distribution of species diversity. *Proc R Soc B Biol Sci.* 2017;284(1855):20170507. https://doi.org/10.1098/rspb.2017.0507.
7. Zaura E, Brandt BW, de Mattos MJT, et al. Same exposure but two radically different responses to antibiotics: resilience of the salivary microbiome versus long-term microbial shifts in feces. *MBio.* 2015;6(6):e01693-15. https://doi.org/10.1128/mBio.01693-15.

8. Dethlefsen L, Relman DA. Incomplete recovery and individualized responses of the human distal gut microbiota to repeated antibiotic perturbation. *Proc Natl Acad Sci U S A*. 2011;108(Suppl. 1):4554–4561. https://doi.org/10.1073/pnas.1000087107.
9. Marais GAB, Calteau A, Tenaillon O. Mutation rate and genome reduction in endosymbiotic and free-living bacteria. *Genetica*. 2008;134(2):205–210. https://doi.org/10.1007/s10709-007-9226-6.
10. Zhao S, Lieberman TD, Poyet M, et al. Adaptive evolution within gut microbiomes of healthy people. *Cell Host Microbe*. 2019;25(5):656–667.e8. https://doi.org/10.1016/j.chom.2019.03.007.
11. Huddleston JR. Horizontal gene transfer in the human gastrointestinal tract: potential spread of antibiotic resistance genes. *Infect Drug Resist*. 2014;7:167–176. https://doi.org/10.2147/IDR.S48820.
12. Human Microbiome Project Consortium. Structure, function and diversity of the healthy human microbiome. *Nature*. 2012;486(7402):207–214. https://doi.org/10.1038/nature11234.
13. Sprockett D, Fukami T, Relman DA. Role of priority effects in the early-life assembly of the gut microbiota. *Nat Rev Gastroenterol Hepatol*. 2018;15(4):197–205. https://doi.org/10.1038/nrgastro.2017.173.
14. Friedman ES, Bittinger K, Esipova TV, et al. Microbes vs. chemistry in the origin of the anaerobic gut lumen. *Proc Natl Acad Sci U S A*. 2018;115(16):4170–4175. https://doi.org/10.1073/pnas.1718635115.
15. Donaldson GP, Lee SM, Mazmanian SK. Gut biogeography of the bacterial microbiota. *Nat Rev Microbiol*. 2016;14(1):20–32. https://doi.org/10.1038/nrmicro3552.
16. Johansson MEV, Sjövall H, Hansson GC. The gastrointestinal mucus system in health and disease. *Nat Rev Gastroenterol Hepatol*. 2013;10(6):352–361. https://doi.org/10.1038/nrgastro.2013.35.
17. Tropini C, Earle KA, Huang KC, Sonnenburg JL. The gut microbiome: connecting spatial organization to function. *Cell Host Microbe*. 2017;21(4):433–442. https://doi.org/10.1016/j.chom.2017.03.010.
18. Falony G, Joossens M, Vieira-Silva S, et al. Population-level analysis of gut microbiome variation. *Science*. 2016;352(6285):560–564. https://doi.org/10.1126/science.aad3503.
19. Bäckhed F, Fraser CM, Ringel Y, et al. Defining a healthy human gut microbiome: current concepts, future directions, and clinical applications. *Cell Host Microbe*. 2012;12(5):611–622. https://doi.org/10.1016/j.chom.2012.10.012.
20. Dewhirst FE, Chen T, Izard J, et al. The human oral microbiome. *J Bacteriol*. 2010;192(19):5002–5017. https://doi.org/10.1128/JB.00542-10.
21. Welch JLM, Rossetti BJ, Rieken CW, Dewhirst FE, Borisy GG. Biogeography of a human oral microbiome at the micron scale. *Proc Natl Acad Sci U S A*. 2016;113(6):E791–E800. https://doi.org/10.1073/pnas.1522149113.
22. Ledder RG, Gilbert P, Huws SA, et al. Molecular analysis of the subgingival microbiota in health and disease. *Appl Environ Microbiol*. 2007;73(2):516–523. https://doi.org/10.1128/AEM.01419-06.
23. Sato Y, Yamagishi J, Yamashita R, et al. Inter-individual differences in the oral bacteriome are greater than intra-day fluctuations in individuals. Martinez-Abarca F, ed. *PLoS One*. 2015;10(6):e0131607. https://doi.org/10.1371/journal.pone.0131607.
24. Simón-Soro A, Tomás I, Cabrera-Rubio R, Catalan MD, Nyvad B, Mira A. Microbial geography of the oral cavity. *J Dent Res*. 2013;92(7):616–621. https://doi.org/10.1177/0022034513488119.
25. Haffajee AD, Teles RP, Patel MR, Song X, Yaskell T, Socransky SS. Factors affecting human supragingival biofilm composition. II. Tooth position. *J Periodontal Res*. 2009;44(4):520–528. https://doi.org/10.1111/j.1600-0765.2008.01155.x.

26. Proctor DM, Fukuyama JA, Loomer PM, et al. A spatial gradient of bacterial diversity in the human oral cavity shaped by salivary flow. *Nat Commun.* 2018;9(1):1–10. https://doi.org/10.1038/s41467-018-02900-1.
27. Dawes C, Watanabe S, Biglow-Lecomte P, Dibdin GH. Estimation of the velocity of the salivary film at some different locations in the mouth. *J Dent Res.* 1989;68(11): 1479–1482. https://doi.org/10.1177/00220345890680110201.
28. Almståhl A, Wikström M, Stenberg I, Jakobsson A, Fagerberg-Mohlin B. Oral microbiota associated with hyposalivation of different origins. *Oral Microbiol Immunol.* 2003;18(1):1–8. https://doi.org/10.1034/j.1399-302X.2003.180101.x.
29. Costalonga M, Herzberg MC. The oral microbiome and the immunobiology of periodontal disease and caries. *Immunol Lett.* 2014;162(2):22–38. https://doi.org/10.1016/j.imlet.2014.08.017.
30. Brogden K, Squier C. *Human Oral Mucosa: Development, Structure, and Function.* Wiley-Blackwell; 2011.
31. Mager DL, Ximenez-Fyvie LA, Haffajee AD, Socransky SS. Distribution of selected bacterial species on intraoral surfaces. *J Clin Periodontol.* 2003;30(7):644–654. https://doi.org/10.1034/j.1600-051X.2003.00376.x.
32. Aas JA, Paster BJ, Stokes LN, Olsen I, Dewhirst FE. Defining the normal bacterial flora of the oral cavity. *J Clin Microbiol.* 2005;43(11):5721–5732. https://doi.org/10.1128/JCM.43.11.5721-5732.2005.
33. Proctor DM, Relman DA. The landscape ecology and microbiota of the human nose, mouth, and throat. *Cell Host Microbe.* 2017;21(4):421–432. https://doi.org/10.1016/j.chom.2017.03.011.
34. Tsukamoto Y, Usui M, Yamamoto G, et al. Role of the junctional epithelium in periodontal innate defense and homeostasis. *J Periodontal Res.* 2012;47(6):750–757. https://doi.org/10.1111/j.1600-0765.2012.01490.x.
35. Abusleme L, Moutsopoulos NM. IL-17: overview and role in oral immunity and microbiome. *Oral Dis.* 2017;23(7):854–865. https://doi.org/10.1111/odi.12598.
36. Shang L, Deng D, Buskermolen JK, et al. Multi-species oral biofilm promotes reconstructed human gingiva epithelial barrier function. *Sci Rep.* 2018;8(1):1–12. https://doi.org/10.1038/s41598-018-34390-y.
37. Pei Z, Bini EJ, Yang L, Zhou M, Francois F, Blaser MJ. Bacterial biota in the human distal esophagus. *Proc Natl Acad Sci U S A.* 2004;101(12):4250–4255. https://doi.org/10.1073/pnas.0306398101.
38. Fillon SA, Harris JK, Wagner BD, et al. Novel device to sample the esophageal microbiome-the esophageal string test. *PLoS One.* 2012;7(9):e42938. https://doi.org/10.1371/journal.pone.0042938.
39. Marshall BJ, Warren JR. Unidentified curved bacilli in the stomach of patients with gastritis and peptic ulceration. *Lancet.* 1984;323(8390):1311–1315. https://doi.org/10.1016/S0140-6736(84)91816-6.
40. Bik EM, Eckburg PB, Gill SR, et al. Molecular analysis of the bacterial microbiota in the human stomach. *Proc Natl Acad Sci U S A.* 2006;103(3):732–737. https://doi.org/10.1073/pnas.0506655103.
41. Li X-X, Wong GL-H, To K-F, et al. Bacterial microbiota profiling in gastritis without helicobacter pylori infection or non-steroidal anti-inflammatory drug use. Ahmed N, ed. *PLoS One.* 2009;4(11):e7985. https://doi.org/10.1371/journal.pone.0007985.
42. Engstrand L, Lindberg M. Helicobacter pylori and the gastric microbiota. *Best Pract Res Clin Gastroenterol.* 2013;27(1):39–45. https://doi.org/10.1016/j.bpg.2013.03.016.
43. Nardone G, Compare D. The human gastric microbiota: is it time to rethink the pathogenesis of stomach diseases? *United European Gastroenterol J.* 2015;3(3):255–260. https://doi.org/10.1177/2050640614566846.

44. Sanduleanu S, Jonkers D, De Bruine A, Hameeteman W, Stockbrügger RW. Non-Helicobacter pylori bacterial flora during acid-suppressive therapy: differential findings in gastric juice and gastric mucosa. *Aliment Pharmacol Ther.* 2001;15(3):379–388. https://doi.org/10.1046/j.1365-2036.2001.00888.x.
45. Vesper B, Jawdi A, Altman K, Haines III G, Tao L, Radosevich J. The effect of proton pump inhibitors on the human microbiota. *Curr Drug Metab.* 2009;10(1):84–89. https://doi.org/10.2174/138920009787048392.
46. Graham DY, Malaty HM, Evans DG, Evans DJ, Klein PD, Adam E. Epidemiology of Helicobacter pylori in an asymptomatic population in the United States: effect of age, race, and socioeconomic status. *Gastroenterology.* 1991;100(6):1495–1501. https://doi.org/10.5555/URI:PII:001650859190644Z.
47. Kuipers EJ, Pérez-Pérez GI, Meuwissen SG, Blaser MJ. Helicobacter pylori and atrophic gastritis: importance of the cagA status. *J Natl Cancer Inst.* 1995;87(23):1777–1780. https://doi.org/10.1093/jnci/87.23.1777.
48. Kabir S. The role of interleukin-17 in the Helicobacter pylori induced infection and immunity. *Helicobacter.* 2011;16(1):1–8. https://doi.org/10.1111/j.1523-5378.2010.00812.x.
49. Perry S, de Jong BC, Solnick JV, et al. Infection with Helicobacter pylori is associated with protection against tuberculosis. Pai M, ed. *PLoS One.* 2010;5(1):e8804. https://doi.org/10.1371/journal.pone.0008804.
50. Blaser MJ. Hypothesis: the changing relationships of Helicobacter pylori and humans: implications for health and disease. *J Infect Dis.* 1999;179(6):1523–1530. https://doi.org/10.1086/314785.
51. Johansson MEV, Phillipson M, Petersson J, Velcich A, Holm L, Hansson GC. The inner of the two Muc2 mucin-dependent mucus layers in colon is devoid of bacteria. *Proc Natl Acad Sci U S A.* 2008;105(39):15064–15069. https://doi.org/10.1073/pnas.0803124105.
52. Zoetendal EG, Raes J, Van Den Bogert B, et al. The human small intestinal microbiota is driven by rapid uptake and conversion of simple carbohydrates. *ISME J.* 2012; 6(7):1415–1426. https://doi.org/10.1038/ismej.2011.212.
53. Bevins CL, Salzman NH. Paneth cells, antimicrobial peptides and maintenance of intestinal homeostasis. *Nat Rev Microbiol.* 2011;9(5):356–368. https://doi.org/10.1038/nrmicro2546.
54. Meyer-Hoffert U, Hornef MW, Henriques-Normark B, et al. Secreted enteric antimicrobial activity localises to the mucus surface layer. *Gut.* 2008;57(6):764–771. https://doi.org/10.1136/gut.2007.141481.
55. Peterson DA, McNulty NP, Guruge JL, Gordon JI. IgA response to symbiotic bacteria as a mediator of gut homeostasis. *Cell Host Microbe.* 2007;2(5):328–339. https://doi.org/10.1016/j.chom.2007.09.013.
56. Mowat AM, Agace WW. Regional specialization within the intestinal immune system. *Nat Rev Immunol.* 2014;14(10):667–685. https://doi.org/10.1038/nri3738.
57. Cullen TW, Schofield WB, Barry NA, et al. Antimicrobial peptide resistance mediates resilience of prominent gut commensals during inflammation. *Science.* 2015;347(6218): 170–175. https://doi.org/10.1126/science.1260580.
58. Martinez-Guryn K, Hubert N, Frazier K, et al. Small intestine microbiota regulate host digestive and absorptive adaptive responses to dietary lipids. *Cell Host Microbe.* 2018; 23(4):458–469.e5. https://doi.org/10.1016/j.chom.2018.03.011.
59. O'Hara AM, Shanahan F. The gut flora as a forgotten organ. *EMBO Rep.* 2006; 7(7):688–693. https://doi.org/10.1038/sj.embor.7400731.
60. Angelakis E, Armougom F, Carrière F, et al. A metagenomic investigation of the duodenal microbiota reveals links with obesity. Covasa M, ed. *PLoS One.* 2015;10(9): e0137784. https://doi.org/10.1371/journal.pone.0137784.

61. Watanabe M, Fukiya S, Yokota A. Comprehensive evaluation of the bactericidal activities of free bile acids in the large intestine of humans and rodents. *J Lipid Res*. 2017; 58(6):1143–1152. https://doi.org/10.1194/jlr.M075143.
62. Gérard P. Metabolism of cholesterol and bile acids by the gut microbiota. *Pathogens*. 2013;3(1):14–24. https://doi.org/10.3390/pathogens3010014.
63. Grill JP, Cayuela C, Antoine JM, Schneider F. Isolation and characterization of a Lactobacillus amylovorus mutant depleted in conjugated bile salt hydrolase activity: relation between activity and bile salt resistance. *J Appl Microbiol*. 2000;89(4): 553–563. https://doi.org/10.1046/j.1365-2672.2000.01147.x.
64. Dussurget O, Cabanes D, Dehoux P, et al. Listeria monocytogenes bile salt hydrolase is a PrfA-regulated virulence factor involved in the intestinal and hepatic phases of listeriosis. *Mol Microbiol*. 2002;45(4):1095–1106. https://doi.org/10.1046/j.1365-2958.2002.03080.x.
65. Ridlon JM, Harris SC, Bhowmik S, Kang DJ, Hylemon PB. Consequences of bile salt biotransformations by intestinal bacteria. *Gut Microbes*. 2016;7(1):22–39. https://doi.org/10.1080/19490976.2015.1127483.
66. Light SH, Su L, Rivera-Lugo R, et al. A flavin-based extracellular electron transfer mechanism in diverse Gram-positive bacteria. *Nature*. 2018;562(7725):140–157. https://doi.org/10.1038/s41586-018-0498-z.
67. Thomson ABR, Drozdowski L, Iordache C, et al. Small bowel review: normal physiology, part 1. *Dig Dis Sci*. 2003;48(8):1546–1564. https://doi.org/10.1023/a:1024719925058.
68. Quigley EMM, Abu-Shanab A. Small intestinal bacterial overgrowth. *Infect Dis Clin North Am*. 2010;24(4):943–959. https://doi.org/10.1016/j.idc.2010.07.007.
69. Sundin OH, Mendoza-Ladd A, Zeng M, et al. The human jejunum has an endogenous microbiota that differs from those in the oral cavity and colon. *BMC Microbiol*. 2017; 17(1):160. https://doi.org/10.1186/s12866-017-1059-6.
70. Wang M, Ahrné S, Jeppsson B, Molin G. Comparison of bacterial diversity along the human intestinal tract by direct cloning and sequencing of 16S rRNA genes. *FEMS Microbiol Ecol*. 2005;54(2):219–231. https://doi.org/10.1016/j.femsec.2005.03.012.
71. Dlugosz A, Winckler B, Lundin E, et al. No difference in small bowel microbiota between patients with irritable bowel syndrome and healthy controls. *Sci Rep*. 2014;5:8508. https://doi.org/10.1038/srep08508.
72. Szentkuti L, Lorenz K. The thickness of the mucus layer in different segments of the rat intestine. *Histochem J*. 1995;27(6):466–472. https://doi.org/10.1007/bf00173712.
73. Villmones HC, Haug ES, Ulvestad E, et al. Species level description of the human ileal bacterial microbiota. *Sci Rep*. 2018;8(1):1–9. https://doi.org/10.1038/s41598-018-23198-5.
74. Booijink CCGM, El-Aidy S, Rajilić-Stojanović M, et al. High temporal and interindividual variation detected in the human ileal microbiota. *Environ Microbiol*. 2010;12(12):3213–3227. https://doi.org/10.1111/j.1462-2920.2010.02294.x.
75. Martinez-Guryn K, Leone V, Chang EB. Regional diversity of the gastrointestinal microbiome. *Cell Host Microbe*. 2019;26(3):314–324. https://doi.org/10.1016/j.chom.2019.08.011.
76. Hooper LV, Littman DR, Macpherson AJ. Interactions between the microbiota and the immune system. *Science*. 2012;336(6086):1268–1273. https://doi.org/10.1126/science.1223490.
77. Ivanov II, Atarashi K, Manel N, et al. Induction of intestinal Th17 cells by segmented filamentous bacteria. *Cell*. 2009;139(3):485–498. https://doi.org/10.1016/j.cell.2009.09.033.
78. Stockinger B, Omenetti S. The dichotomous nature of T helper 17 cells. *Nat Rev Immunol*. 2017;17(9):535–544. https://doi.org/10.1038/nri.2017.50.

79. Huang W, Ramsey KM, Marcheva B, Bass J. Circadian rhythms, sleep, and metabolism. *J Clin Invest.* 2011;121(6):2133–2141. https://doi.org/10.1172/JCI46043.
80. Turek FW, Joshu C, Kohsaka A, et al. Obesity and metabolic syndrome in circadian Clock mutant mice. *Science.* 2005;308(5724):1043–1045. https://doi.org/10.1126/science.1108750.
81. Leone V, Gibbons SM, Martinez K, et al. Effects of diurnal variation of gut microbes and high-fat feeding on host circadian clock function and metabolism. *Cell Host Microbe.* 2015;17(5):681–689. https://doi.org/10.1016/j.chom.2015.03.006.
82. Thaiss CA, Zeevi D, Levy M, et al. Transkingdom control of microbiota diurnal oscillations promotes metabolic homeostasis. *Cell.* 2014;159(3):514–529. https://doi.org/10.1016/j.cell.2014.09.048.
83. Wang Y, Kuang Z, Yu X, Ruhn KA, Kubo M, Hooper LV. The intestinal microbiota regulates body composition through NFIL3 and the circadian clock. *Science.* 2017;357(6354):912–916. https://doi.org/10.1126/science.aan0677.
84. Metcalf AM, Phillips SF, Zinsmeister AR, MacCarty RL, Beart RW, Wolff BG. Simplified assessment of segmental colonic transit. *Gastroenterology.* 1987;92(1):40–47. https://doi.org/10.1016/0016-5085(87)90837-7.
85. Chesson P. Mechanisms of maintenance of species diversity. *Annu Rev Ecol Syst.* 2000;31(1):343–366. https://doi.org/10.1146/annurev.ecolsys.31.1.343.
86. Round JL, Mazmanian SK. Inducible Foxp3$^+$ regulatory T-cell development by a commensal bacterium of the intestinal microbiota. *Proc Natl Acad Sci U S A.* 2010;107(27):12204–12209. https://doi.org/10.1073/pnas.0909122107.
87. Sommer F, Bäckhed F. Know your neighbor: microbiota and host epithelial cells interact locally to control intestinal function and physiology. *Bioessays.* 2016;38(5):455–464. https://doi.org/10.1002/bies.201500151.
88. Bergman EN. Energy contributions of volatile fatty acids from the gastrointestinal tract in various species. *Physiol Rev.* 1990;70(2):567–590. https://doi.org/10.1152/physrev.1990.70.2.567.
89. Donohoe DR, Wali A, Brylawski BP, Bultman SJ. Microbial regulation of glucose metabolism and cell-cycle progression in mammalian colonocytes. Bereswill S, ed. *PLoS One.* 2012;7(9):e46589. https://doi.org/10.1371/journal.pone.0046589.
90. Hamer HM, Jokners D, Venema K, Vanhoutvin S, Troost FJ, Brummer R-J. Review article: the role of butyrate on colonic function. *Aliment Pharmacol Ther.* 2007;27(2):104–119. https://doi.org/10.1111/j.1365-2036.2007.03562.x.
91. Kelly CJ, Zheng L, Campbell EL, et al. Crosstalk between microbiota-derived short-chain fatty acids and intestinal epithelial HIF augments tissue barrier function. *Cell Host Microbe.* 2015;17(5):662–671. https://doi.org/10.1016/j.chom.2015.03.005.
92. Rivera-Chávez F, Zhang LF, Faber F, et al. Depletion of butyrate-producing clostridia from the gut microbiota drives an aerobic luminal expansion of salmonella. *Cell Host Microbe.* 2016;19(4):443–454. https://doi.org/10.1016/j.chom.2016.03.004.
93. David LA, Maurice CF, Carmody RN, et al. Diet rapidly and reproducibly alters the human gut microbiome. *Nature.* 2014;505(7484):559–563. https://doi.org/10.1038/nature12820.
94. Turnbaugh PJ, Bäckhed F, Fulton L, Gordon JI. Diet-induced obesity is linked to marked but reversible alterations in the mouse distal gut microbiome. *Cell Host Microbe.* 2008;3(4):213–223. https://doi.org/10.1016/j.chom.2008.02.015.
95. Sonnenburg ED, Zheng H, Joglekar P, et al. Specificity of polysaccharide use in intestinal bacteroides species determines diet-induced microbiota alterations. *Cell.* 2010;141(7):1241–1252. https://doi.org/10.1016/j.cell.2010.05.005.
96. Kashyap PC, Marcobal A, Ursell LK, et al. Genetically dictated change in host mucus carbohydrate landscape exerts a diet-dependent effect on the gut microbiota. *Proc Natl*

Acad Sci U S A. 2013;110(42):17059–17064. https://doi.org/10.1073/pnas.1306070110.
97. Larsson JMH, Karlsson H, Sjövall H, Hansson GC. A complex, but uniform O-glycosylation of the human MUC2 mucin from colonic biopsies analyzed by nanoLC/MSn. *Glycobiology*. 2009;19(7):756–766. https://doi.org/10.1093/glycob/cwp048.
98. Png CW, Lindén SK, Gilshenan KS, et al. Mucolytic bacteria with increased prevalence in IBD mucosa augment in vitro utilization of mucin by other bacteria. *Am J Gastroenterol*. 2010;105(11):2420–2428. https://doi.org/10.1038/ajg.2010.281.
99. Swidsinski A, Sydora BC, Doerffel Y, et al. Viscosity gradient within the mucus layer determines the mucosal barrier function and the spatial organization of the intestinal microbiota. *Inflamm Bowel Dis*. 2007;13(8):963–970. https://doi.org/10.1002/ibd.20163.
100. Albenberg L, Esipova TV, Judge CP, et al. Correlation between intraluminal oxygen gradient and radial partitioning of intestinal microbiota. *Gastroenterology*. 2014;147(5):1055–1063.e8. https://doi.org/10.1053/j.gastro.2014.07.020.
101. Eckburg PB, Bik EM, Bernstein CN, et al. Microbiology: diversity of the human intestinal microbial flora. *Science*. 2005;308(5728):1635–1638. https://doi.org/10.1126/science.1110591.
102. Pédron T, Mulet C, Dauga C, et al. A crypt-specific core microbiota resides in the mouse colon. *MBio*. 2012;3(3):e00116-12. https://doi.org/10.1128/mBio.00116-12.
103. Nava GM, Friedrichsen HJ, Stappenbeck TS. Spatial organization of intestinal microbiota in the mouse ascending colon. *ISME J*. 2011;5(4):627–638. https://doi.org/10.1038/ismej.2010.161.
104. Wang Y, Devkota S, Musch MW, et al. Regional mucosa-associated microbiota determine physiological expression of TLR2 and TLR4 in murine colon. Yang C-H, ed. *PLoS One*. 2010;5(10):e13607. https://doi.org/10.1371/journal.pone.0013607.
105. Ridlon JM, Kang DJ, Hylemon PB. Bile salt biotransformations by human intestinal bacteria. *J Lipid Res*. 2006;47(2):241–259. https://doi.org/10.1194/jlr.R500013-JLR200.
106. Begley M, Gahan CGM, Hill C. The interaction between bacteria and bile. *FEMS Microbiol Rev*. 2005;29(4):625–651. https://doi.org/10.1016/j.femsre.2004.09.003.
107. Islam KBMS, Fukiya S, Hagio M, et al. Bile acid is a host factor that regulates the composition of the cecal microbiota in rats. *Gastroenterology*. 2011;141(5):1773–1781. https://doi.org/10.1053/j.gastro.2011.07.046.
108. Devkota S, Wang Y, Musch MW, et al. Dietary-fat-induced taurocholic acid promotes pathobiont expansion and colitis in $Il10^{-/-}$ mice. *Nature*. 2012;487(7405):104–108. https://doi.org/10.1038/nature11225.
109. Makishima M, Okamoto AY, Repa JJ, et al. Identification of a nuclear receptor for bite acids. *Science*. 1999;284(5418):1362–1365. https://doi.org/10.1126/science.284.5418.1362.
110. Kawamata Y, Fujii R, Hosoya M, et al. A G protein-coupled receptor responsive to bile acids. *J Biol Chem*. 2003;278(11):9435–9440. https://doi.org/10.1074/jbc.M209706200.
111. Li T, Chiang JYL. Bile acids as metabolic regulators. *Curr Opin Gastroenterol*. 2015;31(2):159–165. https://doi.org/10.1097/MOG.0000000000000156.
112. Sinha SR, Haileselassie Y, Nguyen LP, et al. Dysbiosis-induced secondary bile acid deficiency promotes intestinal inflammation. *Cell Host Microbe*. 2020;27(4):659–670.e5. https://doi.org/10.1016/j.chom.2020.01.021.
113. Round JL, Lee SM, Li J, et al. The toll-like receptor 2 pathway establishes colonization by a commensal of the human microbiota. *Science*. 2011;332(6032):974–977. https://doi.org/10.1126/science.1206095.

114. Halfvarson J, Brislawn CJ, Lamendella R, et al. Dynamics of the human gut microbiome in inflammatory bowel disease. *Nat Microbiol.* 2017;2(5):1–7. https://doi.org/10.1038/nmicrobiol.2017.4.
115. Sokol H, Pigneur B, Watterlot L, et al. Faecalibacterium prausnitzii is an anti-inflammatory commensal bacterium identified by gut microbiota analysis of Crohn disease patients. *Proc Natl Acad Sci U S A.* 2008;105(43):16731–16736. https://doi.org/10.1073/pnas.0804812105.
116. Willing B, Halfvarson J, Dicksved J, et al. Twin studies reveal specific imbalances in the mucosa-associated microbiota of patients with ileal Crohn's disease. *Inflamm Bowel Dis.* 2009;15(5):653–660. https://doi.org/10.1002/ibd.20783.
117. Rajca S, Grondin V, Louis E, et al. Alterations in the intestinal microbiome (Dysbiosis) as a predictor of relapse after infliximab withdrawal in Crohn's disease. *Inflamm Bowel Dis.* 2014;20(6):978–986. https://doi.org/10.1097/MIB.0000000000000036.
118. MacArthur RH, Wilson EO. *The Theory of Island Biogeography*. Princeton University Press; 2001.
119. Wilkinson DM. The disturbing history of intermediate disturbance. *Oikos.* 1999; 84(1):145. https://doi.org/10.2307/3546874.
120. Caporaso JG, Lauber CL, Costello EK, et al. Moving pictures of the human microbiome. *Genome Biol.* 2011;12(5):R50. https://doi.org/10.1186/gb-2011-12-5-r50.
121. Costello EK, Lauber CL, Hamady M, Fierer N, Gordon JI, Knight R. Bacterial community variation in human body habitats across space and time. *Science.* 2009; 326(5960):1694–1697. https://doi.org/10.1126/science.1177486.
122. Grice EA, Segre JA. The skin microbiome. *Nat Rev Microbiol.* 2011;9(4):244–253. https://doi.org/10.1038/nrmicro2537.
123. Sender R, Fuchs S, Milo R. Revised estimates for the number of human and bacteria cells in the body. *PLoS Biol.* 2016;14(8):e1002533. https://doi.org/10.1371/journal.pbio.1002533.
124. Leeming JP, Holland KT, Cunliffe WJ. The microbial ecology of pilosebaceous units isolated from human skin. *J Gen Microbiol.* 1984;130(4):803–807. https://doi.org/10.1099/00221287-130-4-803.
125. Decréau RA, Marson CM, Smith KE, Behan JM. Production of malodorous steroids from androsta-5,16-dienes and androsta-4,16-dienes by Corynebacteria and other human axillary bacteria. *J Steroid Biochem Mol Biol.* 2003;87(4–5):327–336. https://doi.org/10.1016/j.jsbmb.2003.09.005.
126. Natsch A, Gfeller H, Gygax P, Schmid J, Acuna G. A specific bacterial aminoacylase cleaves odorant precursors secreted in the human axilla. *J Biol Chem.* 2003; 278(8):5718–5727. https://doi.org/10.1074/jbc.M210142200.
127. Emter R, Natsch A. The sequential action of a dipeptidase and a β-lyase is required for the release of the human body odorant 3-methyl-3-sulfanylhexan-1-ol from a secreted Cys-Gly-(S) conjugate by Corynebacteria. *J Biol Chem.* 2008;283(30):20645–20652. https://doi.org/10.1074/jbc.M800730200.
128. Grice EA, Kong HH, Conlan S, et al. Topographical and temporal diversity of the human skin microbiome. *Science.* 2009;324(5931):1190–1192. https://doi.org/10.1126/science.1171700.
129. Brüggemann H, Henne A, Hoster F, et al. The complete genome sequence of Propionibacterium acnes, a commensal of human skin. *Science.* 2004;305(5684):671–673. https://doi.org/10.1126/science.1100330.
130. Marples RR, Downing DT, Kligman AM. Control of free fatty acids in human surface lipids by Corynebacterium acnes. *J Invest Dermatol.* 1971;56(2):127–131. https://doi.org/10.1111/1523-1747.ep12260695.
131. Roth RR, James WD. Microbial ecology of the skin. *Annu Rev Microbiol.* 1988; 42(1):441–464. https://doi.org/10.1146/annurev.mi.42.100188.002301.

132. Elias PM. The skin barrier as an innate immune element. *Semin Immunopathol.* 2007; 29(1):3–14. https://doi.org/10.1007/s00281-007-0060-9.
133. Jo JH, Kennedy EA, Kong HH. Topographical and physiological differences of the skin mycobiome in health and disease. *Virulence.* 2017;8(3):324–333. https://doi.org/10.1080/21505594.2016.1249093.
134. Marples MJ. *The Ecology of the Human Skin.* Springfield, IL: Charles C. Thomas; 1965.
135. Gao Z, Perez-Perez GI, Chen Y, Blaser MJ. Quantitation of major human cutaneous bacterial and fungal populations. *J Clin Microbiol.* 2010;48(10):3575–3581. https://doi.org/10.1128/JCM.00597-10.
136. Sumikawa Y, Asada H, Hoshino K, et al. Induction of β-defensin 3 in keratinocytes stimulated by bacterial lipopeptides through toll-like receptor 2. *Microbes Infect.* 2006;8(6):1513–1521. https://doi.org/10.1016/j.micinf.2006.01.008.
137. Nagy I, Pivarcsi A, Kis K, et al. Propionibacterium acnes and lipopolysaccharide induce the expression of antimicrobial peptides and proinflammatory cytokines/chemokines in human sebocytes. *Microbes Infect.* 2006;8(8):2195–2205. https://doi.org/10.1016/j.micinf.2006.04.001.
138. Chehoud C, Rafail S, Tyldsley AS, Seykora JT, Lambris JD, Grice EA. Complement modulates the cutaneous microbiome and inflammatory milieu. *Proc Natl Acad Sci U S A.* 2013;110(37):15061–15066. https://doi.org/10.1073/pnas.1307855110.
139. Naik S, Bouladoux N, Wilhelm C, et al. Compartmentalized control of skin immunity by resident commensals. *Science.* 2012;337(6098):1115–1119. https://doi.org/10.1126/science.1225152.
140. Belkaid Y, Segre JA. Dialogue between skin microbiota and immunity. *Science.* 2014;346(6212):954–959. https://doi.org/10.1126/science.1260144.
141. Cogen AL, Yamasaki K, Muto J, et al. Staphylococcus epidermidis antimicrobial δ-toxin (phenol-soluble modulin-γ) cooperates with host antimicrobial peptides to kill group a streptococcus. DeLeo FR, ed. *PLoS One.* 2010;5(1):e8557. https://doi.org/10.1371/journal.pone.0008557.
142. Anderson DJ, Marathe J, Pudney J. The structure of the human vaginal stratum corneum and its role in immune defense. *Am J Reprod Immunol.* 2014;71(6):618–623. https://doi.org/10.1111/aji.12230.
143. Boskey ER, Cone RA, Whaley KJ, Moench TR. Origins of vaginal acidity: high D/L lactate ratio is consistent with bacteria being the primary source. *Hum Reprod.* 2001; 16(9):1809–1813. https://doi.org/10.1093/humrep/16.9.1809.
144. Ravel J, Gajer P, Abdo Z, et al. Vaginal microbiome of reproductive-age women. *Proc Natl Acad Sci U S A.* 2011;108(Suppl. 1):4680–4687. https://doi.org/10.1073/pnas.1002611107.
145. Hyman RW, Fukushima M, Jiang H, et al. Diversity of the vaginal microbiome correlates with preterm birth. *Reprod Sci.* 2014;21(1):32–40. https://doi.org/10.1177/1933719113488838.
146. Witkin SS. The vaginal microbiome, vaginal anti-microbial defence mechanisms and the clinical challenge of reducing infection-related preterm birth. *BJOG.* 2015; 122(2):213–218. https://doi.org/10.1111/1471-0528.13115.
147. Onderdonk AB, Delaney ML, Fichorova RN. The human microbiome during bacterial vaginosis. *Clin Microbiol Rev.* 2016;29(2):223–238. https://doi.org/10.1128/CMR.00075-15.
148. Donders GGG, Bosmans E, Dekeersmaecker A, Vereecken A, Van Bulck B, Spitz B. Pathogenesis of abnormal vaginal bacterial flora. *Am J Obstet Gynecol.* 2000; 182(4):872–878. https://doi.org/10.1016/S0002-9378(00)70338-3.
149. Serrano MG, Parikh HI, Brooks JP, et al. Racioethnic diversity in the dynamics of the vaginal microbiome during pregnancy. *Nat Med.* 2019;25(6):1001–1011. https://doi.org/10.1038/s41591-019-0465-8.

150. Zhou X, Bent SJ, Schneider MG, Davis CC, Islam MR, Forney LJ. Characterization of vaginal microbial communities in adult healthy women using cultivation-independent methods. *Microbiology*. 2004;150(8):2565–2573. https://doi.org/10.1099/mic.0.26905-0.
151. Zhou X, Brown CJ, Abdo Z, et al. Differences in the composition of vaginal microbial communities found in healthy Caucasian and black women. *ISME J*. 2007;1(2):121–133. https://doi.org/10.1038/ismej.2007.12.
152. Gong Z, Luna Y, Yu P, Fan H. Lactobacilli inactivate chlamydia trachomatis through lactic acid but not H_2O_2. Yang X, ed. *PLoS One*. 2014;9(9):e107758. https://doi.org/10.1371/journal.pone.0107758.
153. Graver MA, Wade JJ. The role of acidification in the inhibition of Neisseria gonorrhoeae by vaginal lactobacilli during anaerobic growth. *Ann Clin Microbiol Antimicrob*. 2011;10:8. https://doi.org/10.1186/1476-0711-10-8.
154. Lai SK, Hida K, Shukair S, et al. Human immunodeficiency virus type 1 is trapped by acidic but not by neutralized human cervicovaginal mucus. *J Virol*. 2009; 83(21):11196–11200. https://doi.org/10.1128/JVI.01899-08.
155. Witkin SS, Mendes-Soares H, Linhares IM, Jayaram A, Ledger WJ, Forney LJ. Influence of vaginal bacteria and D- and L-lactic acid isomers on vaginal extracellular matrix metalloproteinase inducer: implications for protection against upper genital tract infections. *MBio*. 2013;4(4):e00460-13. https://doi.org/10.1128/mBio.00460-13.
156. Hearps A, Gugasyan R, Srbinovski D, et al. Lactic acid, a vaginal microbiota metabolite, elicits an anti-inflammatory response from vaginal and cervical epithelial cells. *AIDS Res Hum Retroviruses*. 2014;30(S1):A238–A239. https://doi.org/10.1089/aid.2014.5527.abstract.
157. Stoyancheva G, Marzotto M, Dellaglio F, Torriani S. Bacteriocin production and gene sequencing analysis from vaginal Lactobacillus strains. *Arch Microbiol*. 2014;196(9): 645–653. https://doi.org/10.1007/s00203-014-1003-1.
158. Aroutcheva A, Gariti D, Simon M, et al. Defense factors of vaginal lactobacilli. *Am J Obstet Gynecol*. 2001;185(2):375–379. https://doi.org/10.1067/mob.2001.115867.
159. Ojala T, Kankainen M, Castro J, et al. Comparative genomics of Lactobacillus crispatus suggests novel mechanisms for the competitive exclusion of Gardnerella vaginalis. *BMC Genomics*. 2014;15(1):1070. https://doi.org/10.1186/1471-2164-15-1070.
160. Nunn KL, Forney LJ. Unraveling the dynamics of the human vaginal microbiome. *Yale J Biol Med*. 2016;89(3):331–337.
161. Mendes-Soares H, Suzuki H, Hickey RJ, Forneya LJ. Comparative functional genomics of Lactobacillus spp. reveals possible mechanisms for specialization of vaginal lactobacilli to their environment. *J Bacteriol*. 2014;196(7):1458–1470. https://doi.org/10.1128/JB.01439-13.
162. Miller EA, Beasley DAE, Dunn RR, Archie EA. Lactobacilli dominance and vaginal pH: why is the human vaginal microbiome unique? *Front Microbiol*. 2016;7(Dec):1936. https://doi.org/10.3389/fmicb.2016.01936.
163. Gajer P, Brotman RM, Bai G, et al. Temporal dynamics of the human vaginal microbiota. *Sci Transl Med*. 2012;4(132):132ra52. https://doi.org/10.1126/scitranslmed.3003605.
164. Chaban B, Links MG, Jayaprakash TP, et al. Characterization of the vaginal microbiota of healthy Canadian women through the menstrual cycle. *Microbiome*. 2014;2(1):23. https://doi.org/10.1186/2049-2618-2-23.
165. Noyes N, Cho KC, Ravel J, Forney LJ, Abdo Z. Associations between sexual habits, menstrual hygiene practices, demographics and the vaginal microbiome as revealed by Bayesian network analysis. *PLoS One*. 2018;13(1):e0191625. https://doi.org/10.1371/journal.pone.0191625.

166. Romero R, Hassan SS, Gajer P, et al. The composition and stability of the vaginal microbiota of normal pregnant women is different from that of non-pregnant women. *Microbiome.* 2014;2(1):1–19. https://doi.org/10.1186/2049-2618-2-4.
167. DiGiulio DB, Callahan BJ, McMurdie PJ, et al. Temporal and spatial variation of the human microbiota during pregnancy. *Proc Natl Acad Sci U S A.* 2015;112 (35):11060–11065. https://doi.org/10.1073/pnas.1502875112.
168. Frank DN, Feazel LM, Bessesen MT, Price CS, Janoff EN, Pace NR. The human nasal microbiota and Staphylococcus aureus carriage. Aziz RK, ed. *PLoS One.* 2010;5(5): e10598. https://doi.org/10.1371/journal.pone.0010598.
169. Yan M, Pamp SJ, Fukuyama J, et al. Nasal microenvironments and interspecific interactions influence nasal microbiota complexity and S. aureus carriage. *Cell Host Microbe.* 2013;14(6):631–640. https://doi.org/10.1016/j.chom.2013.11.005.
170. Bassis CM, Tang AL, Young VB, Pynnonen MA. The nasal cavity microbiota of healthy adults. *Microbiome.* 2014;2(1):27. https://doi.org/10.1186/2049-2618-2-27.
171. Fahy JV, Dickey BF. Medical progress: airway mucus function and dysfunction. *N Engl J Med.* 2010;363(23):2233–2247. https://doi.org/10.1056/NEJMra0910061.
172. Dickson RP, Erb-Downward JR, Martinez FJ, Huffnagle GB. The microbiome and the respiratory tract. *Annu Rev Physiol.* 2016;78(1):481–504. https://doi.org/10.1146/annurev-physiol-021115-105238.
173. West JB. Regional differences in the lung. *Chest.* 1978;74(4):426–437. https://doi.org/10.1016/S0012-3692(15)37392-X.
174. Ingenito EP, Solway J, McFadden ER, et al. Indirect assessment of mucosal surface temperatures in the airways: theory and tests. *J Appl Physiol.* 1987;63(5):2075–2083. https://doi.org/10.1152/jappl.1987.63.5.2075.
175. Charlson ES, Bittinger K, Haas AR, et al. Topographical continuity of bacterial populations in the healthy human respiratory tract. *Am J Respir Crit Care Med.* 2011;184(8):957–963. https://doi.org/10.1164/rccm.201104-0655OC.
176. Dickson RP, Erb-Downward JR, Freeman CM, et al. Spatial variation in the healthy human lung microbiome and the adapted island model of lung biogeography. *Ann Am Thorac Soc.* 2015;12(6):821–830. https://doi.org/10.1513/AnnalsATS.201501-029OC.
177. Venkataraman A, Bassis CM, Beck JM, et al. Application of a neutral community model to assess structuring of the human lung microbiome. *MBio.* 2015;6(1):e02284-14. https://doi.org/10.1128/mBio.02284-14.
178. Erb-Downward JR, Thompson DL, Han MK, et al. Analysis of the lung microbiome in the "healthy" smoker and in COPD. Bereswill S, ed *PLoS One.* 2011;6(2). https://doi.org/10.1371/journal.pone.0016384.
179. Willner D, Haynes MR, Furlan M, et al. Spatial distribution of microbial communities in the cystic fibrosis lung. *ISME J.* 2012;6(2):471–474. https://doi.org/10.1038/ismej.2011.104.
180. Worlitzsch D, Tarran R, Ulrich M, et al. Effects of reduced mucus oxygen concentration in airway Pseudomonas infections of cystic fibrosis patients. *J Clin Invest.* 2002; 109(3):317–325. https://doi.org/10.1172/jci13870.
181. Schmidt A, Belaaouaj A, Bissinger R, et al. Neutrophil elastase-mediated increase in airway temperature during inflammation. *J Cyst Fibros.* 2014;13(6):623–631. https://doi.org/10.1016/j.jcf.2014.03.004.

CHAPTER TWO

The microbiota-gut-brain axis: Focus on the fundamental communication pathways

Michela Bistoletti, Annalisa Bosi, Davide Banfi, Cristina Giaroni*, and Andreina Baj
Department of Medicine and Surgery, University of Insubria, Varese, Italy
*Corresponding author: e-mail address: cristina.giaroni@uninsubria.it

Contents

Abstract

The existence of a bidirectional communication system termed the "gut-brain axis" linking the gastrointestinal tract with the brain has been recognized since the nineteenth century. In the last decades, however, it has become evident that the saprophytic gut microbial flora plays a key role in the modulation of this gut-brain connecting pathway, which is defined now as the "microbiota-gut-brain axis". The gut microbiota plays, indeed, a pivotal role in the maintenance of local, systemic and brain homeostasis. Several neuroactive molecules, hormones and metabolites act along different

Progress in Molecular Biology and Translational Science, Volume 176
ISSN 1877-1173
https://doi.org/10.1016/bs.pmbts.2020.08.012

communication pathways, allowing the bidirectional cross-talk. Some of these molecules are common to eukaryota and prokaryota, representing a possible interkingdom communication system. This chapter provides a description of the chief endocrine, immune and neuronal pathways involved in this bidirectional communication. Research in this field opens an exciting scenario on the possibility to target specific neuronal or hormonal pathways, by means of traditional pharmacological approaches as well as using neuroactive molecule-producing probiotics, as new potential therapeutic tools, addressed to the treatment of neurogastrointestinal and/or psychiatric disorders.

1. Introduction

The sentence "death sits in the bowel", putatively attributed to Hippocrates, the Father of modern Medicine (400 BCE), is indicative of the fundamental role played by the gastrointestinal tract in maintaining the homeostasis of the human body. The gut is essential for the absorption of nutrients but also of several other molecules, which allow the communication with other organs in the body, including the brain. In the nineteenth century, this evidence led to the definition of a bidirectional communication system termed the "gut-brain axis."[1,2] This consists of a complex network of pathways interconnecting the gut and the brain, and encompasses the central nervous system (CNS), the sympathetic and parasympathetic branches of the autonomic nervous system (ANS), the enteric nervous system (ENS), the neuroendocrine and neuroimmune systems.[1,2] It is now clear that motor, sensory, and secretory functions of the gastrointestinal tract are under control of signals generated within the CNS, and conversely, visceral messages from the gastrointestinal tract can modulate brain functions influencing different manifestations of human behavior, both in normal and pathological states.[2–5] Several neuronal, hormonal and immune signaling pathways participate in the communication between the gut and the brain. Some of these molecular players are generated by the intestinal saprophytic flora, which actively contribute to the gut-brain axis function.[1,6] In view of the importance that has been attributed to the gut microbiota in relationship to host health, in more recent years, the term "microbiota-gut-brain axis" has been proposed.[5–7] In healthy individuals, the normal dominant microbiota is relatively stable and forms a mutually beneficial relationship with the host. Locally, the gut microbiota controls metabolic functions, the immune responses and the defense against pathogenic microorganisms. However, the beneficial actions of the commensal microflora extend also to the CNS, and seem to be essential for brain development and

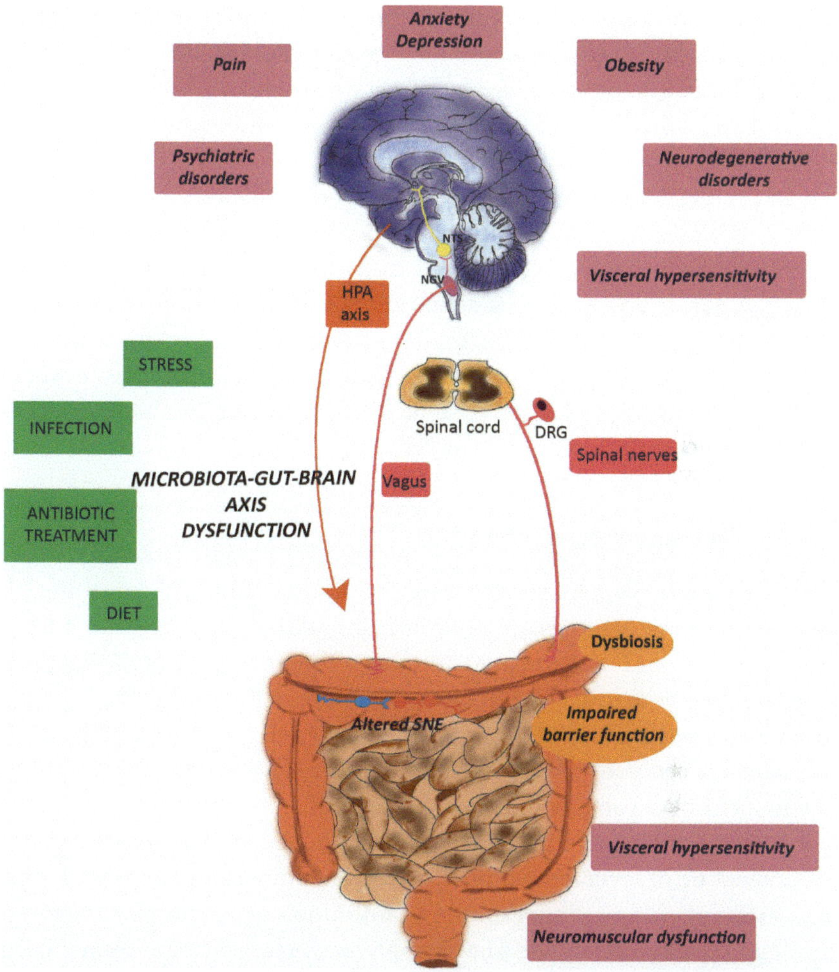

Fig. 1 Several factors, including stress, previous infection, antibiotic treatment and diet may influence the stability of the microbiota–gut–brain axis. Alterations of this bi-directional communication axis are proposed to underlay development of major symptoms involved in the pathogenesis of gastrointestinal diseases, including visceral pain and dysmotility, as well as several CNS disorders. Abbreviations: DRG, dorsal root ganglion; NVG, nodose vagal ganglion; NTS, nucleus of the solitary tract; ENS, enteric nervous system; HPA, hypothalamic-pituitary axis. *Modified from Baj A, Moro E, Bistoletti M, Orlandi V, Crema F, Giaroni C. Glutamatergic signaling along the microbiota-gut-brain axis.* Int J Mol Sci. *2019;20(6). doi:10.3390/ijms20061482.*

homeostasis.[8,9] Alterations in the symbiotic interplay between the microbiota and the host, may, thus, have important consequences, underlying gastrointestinal, metabolic and brain disorders, including neurodevelopmental, neurological and psychiatric diseases[1,4,9–13] (Fig. 1).

An increasing number of neuronal, immune, hormonal and metabolic mechanisms which participate in the control of host homeostasis along the microbiota-gut-brain axis, are being uncovered.[6,9,14,15] However, it seems unlikely that any one of these processes is individually responsible for mediating all microbiota-gut-brain interactions, and these will be discussed individually in subsequent sections of this chapter. Particular attention will be given to the description of neuroactive molecules, hormones and metabolites, which are generated by both eukaryote cells of the host and prokaryotes, giving rise to an interkingdom communication system. Research in this field opens the exciting possibility to target specific neuronal, hormonal and immune pathways in order to discover new potential therapeutic tools, or neuroactive molecule-producing probiotics for the treatment of gastrointestinal and/or brain disorders.

2. The gut microbiota

A rich and dynamic population of microorganisms constitute the enteric microbiota. Bacteria are predominant, with an approximate number of 3.8×10^{13} [16] and an estimated number of genes of about 2–4 million, exceeding greater than 100–150 fold the number of human genes,[16,17] more than 1000 species[17] and 7000 strains.[18] Viruses, Archaea, Fungi and Protozoa also constitute the gut microbiota and, together with bacteria, represent the most abundant microbial population in the human body,[19,20] forming a real organ essential for host homeostasis. As such, the gut microbiota has become the cornerstone of several investigations carried out within the field of nutrition, gastroenterology, endocrinology, immunology, neuropsychiatry and neurology.[6,7] The role of the gut saprophytic microbiota and its symbiotic relationship with the host has been extensively studied in the last few decades by resorting to germ free (GF), dysbiotic animal models and, more recently, by application of metagenomics sequencing techniques.[6]

The gut microbiota undergoes important modifications throughout the lifespan, changing composition, diversity and abundance both in the state of health and disease.[21] Although still debatable,[22] in mammals the womb is not sterile and some data suggest that infants acquire their gut microbiota during intrauterine life, since the placenta, and, occasionally, the amniotic fluid harbor distinct bacterial populations such as *Firmicutes, Tenericutes, Proteobacteria, Bacteroidetes* and *Fusobacteria*.[23–25] It is, however, acknowledged by numerous preclinical studies that neurodevelopment and behavior of the offspring may be influenced by perinatal exposure to microbial products derived from the maternal microbiota.[26] The newborn comes across bacteria for the first-time during delivery. Infants born via vaginal delivery are colonized with bacteria closely related to the mother's vaginal, intestinal or skin maternal microbiota, mainly *Lactobacillus* and *Prevotella* species.[27] Infants delivered by caesarean section, being deprived of contact with maternal vaginal microbiota, are deficient of strict anaerobes, *E. coli*, *Bacteroides* and *Bifidobacteria*, whereas they harbor mainly skin microbes such as, *Staphylococcus* and *Corynebacterium*.[27,28] The gestational age of delivery may also influence the gut microbiota, and differences in its composition are observed between full-term neonates and preterm, probably attributable to the administration of sterile formula and antibiotics, in the latter group.[29] Such microbiota modifications support the development of feeding intolerance and of necrotizing enterocolitis in premature infants.[30] Indeed, through colostrum the infant gut acquires commensal and potential probiotic bacteria, mainly staphylococci and lactic acid bacteria (LAB)[31] and the breast milk contains antimicrobial molecules and prebiotic oligosaccharides, thought to have a beneficial role for the infant.[32,33] As a result, breast milk-fed infants have a microbiota composed of a high number of bifidobacteria and lactobacilli, whereas formula-fed infants have more enterococci and enterobacteria.[31] Noteworthy, the same bacterial strains were found in both breast milk and infant feces of different mother-infant pairs, suggesting a fundamental role of the human milk for the infant gut bacterial colonization.[31] The gut microbiome may undergo major perturbations in its composition during the first 3 years of life and this period is, indeed, critical for the establishment of a healthy and stable microbiota.[34] Several preclinical studies strengthen the concept that perturbations of the microbial community during childhood and adolescence may have significant impact on the developing individual with perduring effects on the host metabolism, physiology and immune system.[26] During these critical periods, alterations in dynamic microbiota–host relationship can profoundly alter brain–gut signaling, increasing the risk of neurodevelopmental disorders.[35]

In adulthood, the composition of gut microbiota is influenced by many factors: diet, age, gender, genetics, geography, health or disease status, pharmacological treatment, hygienic conditions, life style.[36] The most representative gut bacterial phyla in the adult host are *Firmicutes* and *Bacteroidetes*, while *Proteobacteria, Actinobacteria* and *Verrucomicrobia* are less abundant.[34,37]

If unperturbed, the adult human gut microbiota is stable, but in old age (over 65 years), mainly because of the physiological changes, which increase the susceptibility to disease, its stability deteriorates dramatically changing its composition.[38–41] In particular, elderly people manifest an increased incidence of gastrointestinal disorders, correlated to degenerative changes in intrinsic neuronal circuitries modulating secretion and the motor function. Elderly age-induced gut dysfunctions are associated with profound remodeling of the microbiota, whose hallmark is a marked decreased diversity.[39–41] Notably, while Firmicutes are more abundant versus Bacteroidetes in the adult gut, this ratio appears to be inverted during ageing.[42,43] Interestingly, certain specific taxa are enriched in centenarians, such as *Akkermansia*, suggesting that the presence of specific beneficial bacteria in the gut might be correlated with health in old age.[39] The gut microbiota may also participate to the development of diseases in aged people, such as those associated with an increased inflammatory status and decreased adaptive immunity.[44,45] Specific microbial species with anti-inflammatory properties, such as *Bacteroidetes*, *Clostridium* cluster IVa, and *Faecalibacterium prausnitzii* seem to be associated with reduced age-related fragility, while *Porphyromonadaceae* correlate to cognitive and affective disorders,[46–48] as also demonstrated in preclinical investigations on aged mice.[49] In a clinical study, administration to an elderly cohort of a probiotic, *Lactobacillus rhamnosus GG,* favored the bacterial population reconstitution towards an anti-inflammatory phenotype that promotes the survival and growth of beneficial microbes.[50] Importantly, dietary restriction, which is associated with changes of gut microbiota composition, is a key determinant of healthy aging, representing the most efficacious strategy to increase lifespan in several species.[6] In this view, modulation of gut microbiota composition by pre- and probiotic administration, diet restriction, and, more recently, by fecal microbiota transplantation (FMT) may represent potential innovative therapeutic strategies for aging-associated diseases.[51]

2.1 Microbiota and host interaction: An interkingdom communication

The symbiosis between the gut microbiota and the host is the result of a dynamic and beneficial equilibrium between both players. In this scenario, exponentially growing studies focus the attention on the fascinating interkingdom communication developed between prokaryotic and eukaryotic cells (Fig. 2).

Microorganisms harboring within the human gastrointestinal tract produce many compounds useful to the host, such as vitamins, gas, organic acids, bile salts, bacteriocin.[52,53] Moreover, bacteria influence absorption, distribution and metabolism of nutrients and drugs and may reinforce host

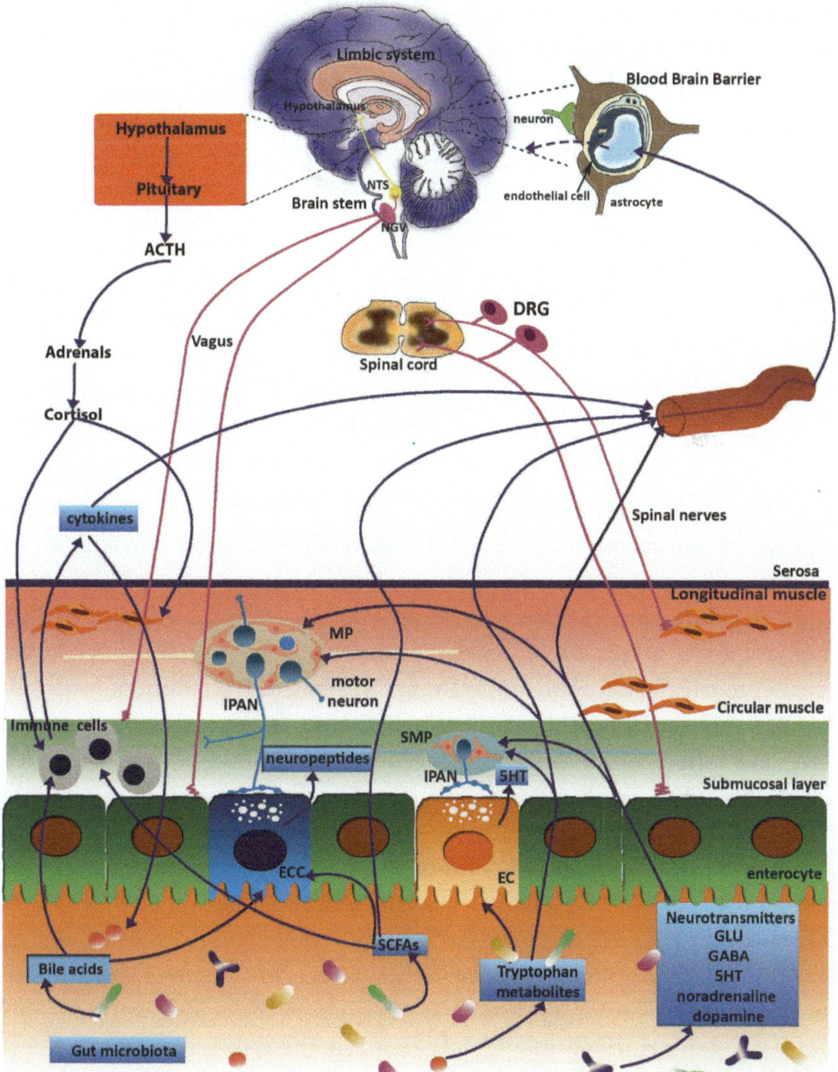

Fig. 2 Schematic representation of the microbiota–gut–brain axis. The gut saprophytic microflora signals to the central nervous system (CNS) and to the enteric nervous system (ENS) via different pathways, including endocrine, immune, metabolic and neuronal pathways explained throughout the text. Abbreviations: NTS, nucleus of the solitary tract; NVG, nodose vagal ganglion; DRG, dorsal root ganglion; MP, myenteric plexus, IPAN, intrinsic primary afferent neurons, SMP, submucosal plexus, ECC, enteroendocrine cell; EC enterochromaffin cells; SCFA, short chain fatty acid. *Adapted from Baj A, Moro E, Bistoletti M, Orlandi V, Crema F, Giaroni C. Glutamatergic signaling along the microbiota-gut-brain axis.* Int J Mol Sci. *2019;20(6). doi:10.3390/ijms20061482.*

innate and acquired immunity, representing a biological barrier against pathogens.[1] GF animals, if compared with conventionally reared controls, necessitate of a higher caloric intake to maintain the same body weight and are prone to vitamin deficiencies requiring dietary supplementation.[54,55] The main mechanisms used by the gut microbiota to maximize nutrient availability are represented by the calories released from otherwise unavailable oligosaccharides and the modulation of absorption.[7] Indeed, a significant source of energy for humans is represented by shorth chain fatty acids (SCFAs), produced from the bacterial metabolism of dietary fibers.[56] Amino acids released by bacteria can also influence host physiology, as they can be used for biosynthetic needs, and their deamination causes the accumulation of ammonia, carbon dioxide and SCFAs.[57] SCFAs and other microbial metabolites, such as polyamine (i.e., putrescine, spermidine, spermine) and aryl hydrocarbon receptor ligands, have also immunomodulatory properties, influencing the host immune response.[58] In addition, many structural bacterial components take part to the cross-talk with the host immune system, such as lipopolysaccharide, lipoteichoic acids, peptidoglycan, flagellin, formyl peptides and unique nucleic acid structures.[58] The overabundance of microbial virulence factors (i.e., pigments, proteases, nuclease, toxins, haemophores), on reverse, represent a family of molecules detrimental to the host health.[59] On the host side, the gut is an optimal nutritional and physical environment, favoring bacteria metabolic activities and growth. Thus, the human gut may be viewed as a "melting pot," where metabolites, cellular components, hormones, virulence factors, are released by prokaryotic and eukaryotic cohorts along a bidirectional communication pathway.[1]

In this mutual interaction, whilst human molecules are well-known and characterized, prokaryotic analogues remain largely to be unveiled. Similarly to eukaryotes, prokaryotes possess an intra- and interspecies communication system called quorum sensing (QS),[60] enabling them to act as a group.[61] QS, which is involved in many aspects of bacterial physiology, comprises small diffusible signal molecules, termed autoinducers (AIs).[62] One of them, AI-2, is shared by both Gram-positive and Gram-negative bacteria and modulates the abundance of the gut microbiota, shaping its structure.[63] QS molecules play an important role in microbiota–host interactions,[64,65] eliciting different responses in mammalian cells,[66] such as those involving activation of inflammatory pathways.[67–69] There are reports suggesting that QS participates to bacterial gut biofilm formation,[70,71] which is involved in the pathogenesis of gut diseases, such chronic inflammatory bowel disease,[60,71–73] and with an increased risk for sporadic colorectal cancer.[72,74] The possibility that signals generated by Als may contribute to gut inflammation and carcinogenesis is, thus, intriguing.[72,74] An important aspect of the interplay between the host and the saprophytic microflora is the co-evolution between the gut microbiota and the host immune system during the host lifespan.[75] In this view, bacterial QS signaling was shown to participate to the host immune responses,[76,77] with both immunomodulatory[78] and immune suppressive activity.[79,80] Thus, it is reasonable to hypothesize the presence of AIs sensing receptors on eukaryotic cells.[62] Conversely, several studies demonstrated that host neurotransmitters and hormones may interact with AIs, thus, influencing gut microbiota homeostasis.[81,82] In the human gut, adrenaline can mimic the activity of AI-3a and its receptor QseC,[83] while noradrenaline promoted the growth of enterohemorragic *E. Coli*, to express its virulence traits.[84,85] Similarly, other neurotransmitter and hormones, such as dynorphine, substance P, calcitonin gene-related peptide, somatostatin, contained in and secreted by enteroendocrine and enterochromaffin cells, immune cells, and nerve endings, have been suggested to interact with bacterial sensor thus impacting in the modulation of bacterial virulence, adhesion, and/or biofilm formation.[70]

3. The interplay between the gut microbiota and the enteric environment

An area of intensive research is developing to uncover the mechanisms underlying the cross-talk occurring among the microbiota and the different cell populations within the enteric microenvironment.[6,86] The latter is represented by a dynamic mosaic of cellular "players" including enterocytes, enteroendocrine cells, neurons, enteric glial cells, smooth muscle cells, the interstitial cells of Cajal, which represent intestinal pacemaker cells, immune cells, which undergo adaptive changes both in physiological and pathological conditions (Fig. 2).[87] The interaction between this vast array of cellular populations and the saprophytic microflora may extend, via the autonomic nervous system (ANS), humoral and hormonal pathways, mainly the hypothalamic pituitary axis (HPA), to the CNS, giving rise to the microbiota-gut-brain communication axis. In the following paragraphs, the bidirectional cross-talk between the microbiota, microbiota metabolites and specific cell populations within the enteric microenvironment as well as the involvement of ascending and descending communication pathways will be described.

3.1 The gut microbiota shapes the gut epithelial barrier structure and function

The gut epithelium represents the widest mucosal surface in the human body, representing an interface between the host and the outer environment and, as such, has a fundamental homeostatic role warranting health status of the organism. The intestinal epithelium, which is composed of a single cell layer, limits the contact of intestinal microbiota with the underlying intestinal tissue, by secreting a protective viscous mucus layer from specialized goblet cells. The interaction between the host and microbes predominantly occurs within this mucosal interface, and serves to enable the communication between the gut and the immune system through antigen recognition, thus activating defensive immune system responses towards pathogens.[88] In addition to the mucus layer, tight-junction proteins on enterocytes (such as occludin, junctional adhesion molecule, and claudin family members that interact with cytoplasmic linker proteins, such as zonula occludens-1, ZO-1) contribute to sustain the barrier function against bacteria and antigens originating from the outer environment.[89] The gut microbiota can regulate the development and function of the intestinal barrier in different ways, such as, by favoring epithelial cell proliferation, enhancing epithelial cell integrity via translocation of tight junction proteins and up-regulating gene expression involved in desmosome maintenance.[90,91] For example, in a seminal study, monoassociation of GF animals with *Bacteroides thetaiotaomicron* up-regulates sprr2a, a gene involved in desmosome maintenance and reinforcement of the intestinal barrier.[92] Furthermore, saprophytic bacteria of the *Bifidobacterium* and *Lactobacillus* species, enhance the survival of gut epithelial cells by inhibiting pro-apoptotic pathways associated with pathogenic bacteria.[93–95] The reduced intestinal surface area, slower epithelial cell turnover, increased enterochromaffin cell area and smaller villous thickness observed in GF animals, with respect to conventionally reared animals, support the hypothesis for a central role played by the gut microbiota in preserving the intestinal barrier architecture and function.[96] Enterocytes can recognize bacterial molecular patterns via a highly conserved family of pathogen-associated molecular pattern (PAMP) receptors, the most studied of which are represented by the Toll-like receptors (TLRs) family, located on their surface.[97] Until now, 13 TLRs have been identified in mammals and are expressed by a variety of cells including not only epithelial cells, but also immunocytes, neurons and glial cells both the in the peripheral and central nervous system. TLRs are further classified into subfamilies, according to substrate that they recognize. TLRs located in the plasma membrane interact with molecules constitutive within the bacterial wall and membrane, such as lipopeptides, peptidoglycans and glycolipids (TLR1/TLR2, TLR2/TLR6, TLR2/TLR10), fibronectin, lipopolysaccharides (LPS), heat shock proteins (TLR4), and flagellin (TLR5). Intracellular TLRs, expressed in cell endosomes, detect microbial nucleic acids, specifically viral double-strand RNAs (TLR3), single-strand RNAs (TLR7 and TLR8), unmethylated CpG DNA (TLR9).[98] Activation of TLRs promotes intracellular signals associated with distinctive adaptor proteins, such as myeloid differentiation primary response gene 88 (MyD88) and TIR- domain-containing adapter-inducing interferon-β (TRIF) pathway. As a consequence, several downstream pathways, comprising the nuclear factor-kappa B (NF-κB), mitogen-activated protein kinases (MAPK), and/or interferon-regulatory factor (IRF) signaling pathways, are activated to induce the expression of a variety of gene products.[98,99] Activation of TLRs by specific bacterial products, for example, induced epithelial cell proliferation, thus favoring epithelial surface repair, following injury.[100] In addition, TLR2 signaling pathway directly reinforces the integrity of enterocytes sustaining the translocation of ZO-1.[101]

A healthy gut microbiota seems essential to allow a normal function to secretory Paneth cells, which cluster in small cell assembly at the base of crypts of Lieberkühn in the epithelium of the small intestine and contribute to intestinal innate immunity by secreting a diverse array of antimicrobial proteins.[102,103] Paneth cells have been shown to directly interact with bacteria metabolites, via TLRs, to express multiple antimicrobial factors which impede commensal and pathogenic bacteria penetration through the epithelial barrier.[103] These anti-microbial peptidergic factors, comprising α-defensins and lysozyme C, and their synthesis require a complete microbiome.[103–105] Interestingly, *Lactobacillus* and *Bifidobacterium* were able in vitro to inhibit Listerial infections by producing antibacterial protein.[106] The same group observed that *Lactobacillus salivarius* was able, in vivo, to produce an antibacteric bacteriocin, which preserved mice against an infection induced by the pathogen *Listeria monocytogenes.*[107]

Under certain pathological conditions, such as during infection or stress, the integrity of the epithelium is, however, compromised and the barrier becomes leaky, thus allowing the translocation of pathological bacteria across the mucosal lining. Deterioration of the barrier integrity may allow the direct interaction of the microorganisms and/or of their by-products with cells in the enteric microenvironment, especially with immune cells and enteric neurons.[108] This leads to activation of an immune response characterized by increased production of proinflammatory mediators which may influence the gut function but may also extend their action to more distal sites, reaching the CNS. Indeed, deranged intestinal permeability has been correlated not only to the gut dysfunction but also to psychiatric disorders, such as depression.[109] For example, increased antibody titers to LPS derived from Gram-negative enterobacteria were found in peripheral blood of patients with major depression with respect to normal volunteers.[109–111] Although, the cause/effect relationship between the occurrence of intestinal permeability changes and brain disorders is still to be clearly defined, the possibility to reconstitute a healthy epithelial barrier with microbiota-based approaches has been suggested as a possible adjuvant treatment of these diseases.[112] In this direction, preclinical studies showed that administration of probiotics such as *Lactobacillus farciminis*, *Bacteroides fragilis*, and *Lactobacillus salivarius* could reverse the leakiness induced by acute stress, maternal infection, and hydrogen peroxide, respectively.[113–115]

3.2 The microbiota and enteroendocrine signaling

Besides providing a physical barrier, the epithelium contains a number of different cell types including enteroendocrine cells (EECs), which represent sensory cells within the intestinal epithelium. ECCs are present as single cells in the crypts and villi of the gastrointestinal mucosa and represent the largest endocrine system of the human body.[116] At least 15 types of EEC have been described, capable of secreting over 20 peptide hormones, which include members of the chromogranin/secretogranin family, serotonin (5-HT), somatostatin, neuropeptide Y (NPY), vasoactive intestinal peptide (VIP), substance P (SP), cholecystokinin (CKK), glucagon-like peptide (GLP)-1/2, and ghrelin.[117,118] EECs have a fundamental role in the maintenance of the human body homeostasis by releasing signaling molecules, which are involved in the regulation of gut motility, gastric acid secretion, metabolic and behavioral responses, such as insulin secretion or food intake.[119] The apical surface of ECCs comes into contact with luminal constituents, including bacterial metabolites, and by releasing peptides and biological active molecules, such as 5-HT, participates to the local signaling network of enteric neurons, glia and immune cells in the intestinal submucosal layer.[120,121] Gut signaling molecules from EECs may influence also the brain function, via hormonal and humoral route, thus participating to the gut-brain communication axis.[59] This interplay extends also to the gut microbiota since the saprophytic microflora has a modulatory role on the release of gut peptides from ECCs.[122]

3.2.1 Enteroendocrine cells

Enteroendocrine L cells secrete anorexigenic hormones involved in the modulation of eating, such as glucagon-like peptide-1 (GLP-1) and peptide YY (PYY) during the postprandial state.[123,124] The main stimulus for postprandial release of GLP-1 and PYY in the proximal gut, is represented by luminal nutrients.[125] Receptors for both hormones are located along the gut-brain axis, in the ENS on vagal afferents and in the brain stem and hypothalamus in the CNS.[126,127] GLP-1 and PYY can stimulate satiation, thus inhibiting eating by acting either directly on the hypothalamic centers of appetite control, or indirectly via the vagal-brainstem-hypothalamic pathway.[125,126] Recent data, however, suggest the occurrence of a peripheral interplay between L cells and gut sensory epithelial cells that synapse with vagal neurons, thus favoring a faster regulation of EECs signaling from the gut to the brain.[128] L cells can sense diverse food components by G protein coupled receptors and transporters located on the apical border,[129] including free fatty acid receptor types (FFAR1–4) for short chain fatty acids (FFAR 2 and 3) and long chain fatty acids (FFAR 1 and 4)[130]; the sodium-coupled glucose transporter (SLC5A1)[131] for carbohydrates; the GPRC6A, Ca^{++} sensing receptor and taste receptors for aminoacids; the GPR119 for oleylethanolamide and GPR93 for peptone.[132] L cells located in more distal gut regions are almost exclusively sensitive to bacteria-derived metabolites, and bacterial products, such as LPS, may enhance GLP-1 secretion.[133] Indeed, it is now ascertained that the gut microbiota may contribute to gut chemosensitivity.[134] Interestingly, activation of TLRs located in a human cell line of L cells induced PYY expression in a NF-kβ-dependent way.[135] Bacterial SCFA metabolites, like butyrate and propionate, by activating FFAR2, have been shown to regulate L type cells density and PYY release in mouse colon.[136] In addition, mice lacking both FFAR2 and FFAR3 exhibited reduced SCFA-triggered GLP-1 secretion in vivo and in vitro and a parallel impairment of glucose tolerance.[137] Activation of the G protein coupled bile acid receptor Gpbar1 receptor (TGR5) on L cells by the bacteria-derived secondary bile acids, lithocholic acid and deoxycholic acid, induced the secretion of both GLP-1 and PYY.[138,139] The ability of bacterial metabolites derived from non-digested nutrients to modulate the activity of colonic L cells is long lasting and proceeds during fasting periods,[140] allowing a basal release of anorexogenic GLP-1 and PYY, which may be beneficial especially for obese individuals, generally characterized by lower serum levels of both GLP-1 and PYY.[126] In this view, the possibility to influence the gut microbiota composition and function with a dietary supplementation of pre- and probiotics in order to modulate glucose tolerance, food intake and decrease body weight, has given promising results in preclinical models of obesity and diabetes. For example, the prebiotics inulin, as well as fructooligosaccharide (FOS) and galactooligosaccharide (GOS) increased the production of both GLP-1 and PYY,[141,142] while certain Lactobacillus strains, induced GLP-1 secretion both in vitro and in vivo.[143–145]

3.2.2 Enterochromaffin cells

Enterochromaffin cells (ECs) are located in the epithelial layer of the entire gastrointestinal tract and, similarly to intestinal epithelial cells, are accessible by microbiota metabolites on the luminal side, while the basolateral border is in contact with afferent and efferent nerve terminals located on the *lamina propria.*[117,118] ECs may, thus, have a peculiar function in the bi-directional transmission of signals from the microbiota to the brain, although, dissimilar to L cells, interactions between gut microbiota and ECs are less clear-cut described.[6] ECs secrete the biogenic amine, serotonin (5-HT) and signaling peptides (for example, corticotrophin releasing hormone, cholecystokinin and somatostatin) in response to various physiological and pathological luminal stimuli,[117,118] including microbial metabolites or bacterial toxins,[146,147] as well as to central stimuli.[148] For example, in conscious rats, intracerebroventricular injection of TRP, an analog of thyrotropin-releasing hormone (TRH), which represents a central mediator of the stress response to cold temperatures, induced a vagally-mediated 5-HT secretion from gastric ECs into the lumen.[148] 5-HT released by ECs activates in a paracrine modality its own receptors on intrinsic and extrinsic primary afferent nerve terminals laying in proximity of ECs, thus leading to activation of enteric reflexes mediating peristalsis, secretion, pain perception and inflammation.[149,150] From a functional view point, ECs are endowed with a vast array of receptors, including 5-HT receptors, pituitary adenylate-cyclase-activating peptide, α- and β-adrenergic, cholinergic, the corticotropin-releasing hormone and γ-aminobutyric acid (GABA) receptors for neurotransmitters and hormones.[151] Intriguingly, a murine ECs cell line, STC-1, expresses various TLR receptors, thus controlling host–microbe signaling.[152] Thus, considering the interkingdom communication system between the host and bacteria metabolites, the expression of multiple receptors could represent a wide array of available targets, through which the microbiota may influence 5-HT release. Secretion of the enterotoxin, cholera toxin, from Vibrio cholerae, induces fluid secretion via GM1 ganglioside receptor activation on ECs.[153]

The presence of pathogenic bacteria in the gut lumen, including *E. coli*, *V. cholerae*, or *Salmonella typhimurium*, has been associated with upregulation of 5-HT secretion into the *lamina propria*.[154] Consequently, paracrine stimulation of intrinsic and extrinsic primary afferents results in local activation of secretomotor reflexes underlying chloride ions and water release within the intestinal lumen.[155] Enhancement of intraluminal fluid content stimulates bowel motility, which, ultimately, favors clearing of intestinal contents, including pathogenic bacteria. Upregulation of mucosal 5-HT has been hypothesized to underlie the development of postinfectious bowel disorders. In these conditions, inflammation-induced upregulation of 5-HT signaling persists after the inflammation has regressed, and, microbe to ECs signaling may result in persistent symptoms of bowel dysfunction in human, such as the post infectious irritable bowel syndrome (IBS).[156,157] It is improbable that variations in intestinal 5-HT released from ECs may exert a direct effect on the CNS since 5-HT cannot cross the blood brain barrier (BBB).[158] However, 5-HT released from ECs can potentially participate to brain-gut communication by modulating vagal afferent activity[159,160] and inflammatory responses in the gut[161] as discussed in a successive paragraph.

3.3 Gut lymphoid associated tissue and the gut microbiota

In addition to neural, endocrine, and metabolic pathways, immunological mechanisms may convey important signals within the microbiota-gut-brain axis.[162,163] The human immune system comprises the innate immunity, that displays similar responses to all adverse agents, and the adaptive immunity, that develops specific responses after recognizing specific antigens and retains memory of the exposure.[163] Since the mucosal gut layer is continuously exposed to antigen stimuli, the defensive function of the gut requires different factors which trigger either the innate and adaptive immune system, via activation of complex set of interplaying mechanisms within the GALT (gut associated lymphoid tissue), the largest lymphoid tissue in the human body.[163] Studies carried out on GF animals revealed that the microbiota is crucial for the GALT development and plays a fundamental role in modeling innate immune responses in the gut.[162,164] The gut saprophytic microflora influences the relative populations, migration, and function of various subsets of immune cells of the innate and adaptive immunity and the mechanisms underlying this modulatory function, in health and disease states, are progressively revealed.[57,162–164]

3.3.1 Innate immune system

Both inductive (Peyer's patches) and effector sites (*lamina propria* and subepithelial cells) constitute the GALT, which acts to prevent the transit of external, potentially harmful, antigens across the epithelial barrier, but is also in constant contact with the microbiota and with food derived antigens. Whilst enterocytes express innate immune receptors and can release cytokines and chemokines, the GALT tissue utilizes lymphocytes to mount a more specific immune response via secretion of polymeric immunoglobulin A (IgA) and modulates oral tolerance through processes that involve specific cytokine and peculiar cell populations.[165] Innate immunity involves the activity of monocytes/macrophages, basophils, eosinophils, and neutrophils, in addition to mast cells, platelets, and natural killer cells, M cells and dendritic cells (DC).[165] DCs of the GALT can sample the microbial milieu while M cells efficiently take up in the lumen and transport towards the *lamina propria* a variety of microorganisms and antigens via active vesicular transport across the epithelium. *Lamina propria* DCs process antigens from intestinal lumen by sending dendrites between enterocytes and present them to T naïve lymphocytes. In this setting, DCs modulate immune responses through activation of signaling events leading to improve expression of factors, such as cytokines and chemokine that recruit and regulate the phenotype and functions of immune T cells.[166,167]

Peptidoglycans, polysaccharides, and other bacterial antigens that have a protective role for bacterial cells (i.e., protection against degradation), also allow the host immune cells to identify numerous and diverse bacteria species. The fundamental mechanism of the innate immunity mucosal system is its ability to distinguish potentially pathogenic microbes from innocuous antigens via activation of PAMPs, principally TLRs. Epithelial TLRs, once activated, can favor the release of inflammatory mediators, the production of cytokines, and chemokine-induced recruitment of inflammatory cells.[168] Interestingly, TLRs are not only engaged by PAMPs, but also by an array of endogenous molecules generated during tissue damage (damage-associated molecular patterns, DAMPs), activating firstly an acute immune response and, secondarily, tuning the subsequent adaptive immune reaction.[97] Many commensal bacteria metabolites including neuromodulators, bacteriocins, bile acids, choline, and SCFAs are also immunomodulatory. Ligands from commensal bacteria such as polysaccharide A, lipopolysaccharide (LPS), and lipoteichoic acid (LTA) influence the normal development and function of the mucosal immune system.[100,169] The commensal microflora seems to be fundamental for a fully developed and functional gut innate immune system. In comparison to conventionally reared animals, GF animals have decreased plasma cells and IgA, decreased expression of activation markers on intestinal macrophages, decreased MHCII on epithelial cells, decreased nitric oxide, and histamine levels in the small intestine.[170–175] In GF animals, Peyer's patch follicles are reduced in number and size and the mesenteric lymph nodes are smaller, with less cellular elements and do not have germinal centers.[170,171] However, reconstitution of GF mice with an intestinal microflora is sufficient to restore the mucosal immune system.[176]

Although the interaction between the gut microbiota and immune cells in the brain has not been investigated hitherto in detail, an emerging body of evidence is recently available to suggest that cytokines and chemokines released in consequence of the microbiota-host interaction can pass into blood and lymphatic vessels, or influence vagal and spinal afferent neuronal pathways, thus communicating with the CNS to regulate brain function and behavior.[6] Indeed, the interaction between metabolites deriving from the gut microbiota and microglia is receiving attention as a key mechanism underlying microbiota-immune-brain interactions.[177] Microglia, the resident innate immune cells in the CNS, are responsible for sensing antigen molecules in the brain microenvironment,[178–180] inducing neuroinflammation in the CNS.[181–183] Microglial cells may undergo plasticity and subsequently release cytokines and chemokines, express numerous antigenic markers, recruit monocytes from the periphery to the brain to aid in defense and cell debris clearing and regulate neurotransmitter pathways.[184,185] Besides their established immune role in the CNS, these cells have a fundamental role in the remodeling of neuronal circuitries during development and adulthood.[184,186] Recently, it was demonstrated that the gut microbiota composition may influence the maturation and the maintenance of a healthy and functional microglia, consequently favoring brain functions both during development and in adult age.[187,188] In both GF and dysbiotic animals, the absence of a complex microbiota favors the development of a hyperplastic microglia, displaying defects in differentiation, morphology and compromised immune response to bacterial or viral infection.[187] Some of these alterations are reversed with recolonization of gut microbiota, following co-habitation of GF mice with control mice.[187] Interestingly, SCFAs may favor the stability of a healthy microglia density, morphology, and maturity in the absence of a complex microbiota,[187] thus suggesting that the GI microbiota may indirectly influence the brain function through regulation of monocyte trafficking to the CNS and subsequent microglia activation, possibly via SCFA-mediated pathways.

3.3.2 Adaptive immune system

Adaptive immunity comprises cells originating in the lymphoid lineage (i.e., B and T lymphocytes) which are able to initiate specialized responses to pathogens characterized by specificity, memory, regulation, diversity, and tolerance.[189] Furthermore, these cells are also involved in inflammatory responses by releasing both pro- and anti-inflammatory cytokines.[190] Oral tolerance, for example, represents a mechanism impeding the development of inflammatory responses against food proteins and against the resident intestinal bacterial microbiota. This is accomplished by activation of a tolerogenic mechanism involving $CD4^+$ T cell naïve, that suppress the expression of T effector cells (T helper 1, Th1, that modulate cell-mediated immunity by secreting INFγ and TNFα and T helper 2, Th2, that modulate humoral immunity by secreting IL-4 and IL-6) and stimulate the expansion of regulatory T cells (Tregs), via secretion of cytokines such as IL-10 and TGFβ.[189,191] Antigens are presented by dendritic cells in the contest of MHC class II molecules to naïve T lymphocytes and, in healthy conditions the release of cytokines from naïve T lymphocytes is tightly regulated. In this scenario, there are several reports suggesting that the intestinal microflora and its qualitative differences in composition might affect adaptive immunological homeostasis.[191] For example, different probiotics such as *Lactobacillus* and *Bifidobacterium* displayed in vivo and in vitro immunomodulatory effects including control of Th1/Th2 balance and cytokine expression, whose alteration may induce disease states.[192,193] Microbial colonization also influences TLRs on gut immune cells that, by recognizing PAMPs may modulate also the adaptive immune response, according to the characteristics of the microbial strain (commensal or pathogen).[194] Recognition of pathogenic antigens by specific TLRs is followed by intracellular activation of the NF-κB signaling system which results in cytokine production and T cell activation.[195] It must be pointed out, however, that in the absence of the resident microbiota, the expression levels of some TLRs are low or absent, thus hampering proper immune responses to pathogenic stimuli.[196] The balance between microbiota, immune response and tolerance mechanisms is fundamental not only for a healthy intestine but may also affect the bidirectional communication with the CNS making it an important target for transducing the effects of bacteria to the CNS.[6] Indeed, the potential role the adaptive immune system in the modulation of brain and behavior has been deeply investigated in the last few years.[197] In addition, the adaptive immune system develops postnatally and is, therefore, critically shaped by exposure to microbes, suggesting that development of some CNS disorders may be depend upon the disruption of microbes and immune function.[6,197,198]

In recombination activation gene 2 knockout (Rag2) transgenic mice, deficient in mature B and T lymphocytes, the adoptive transfer of lymphocytes from stressed mice ameliorated social behavior and reduced anxiety and enhanced hippocampal cell proliferation and oriented microglia towards an anti-inflammatory, neuroprotective phenotype compared with those receiving no cells or cells from unstressed donors.[199] These findings reveal that adaptive immune cells putatively involved in the mechanisms underlying behavioral response to stress and a potential target for developing novel rapid-acting antidepressant therapies.[199] The observation that, in $Rag2^{-/-}$, adaptive immunity influences microbiota composition, supports the hypothesis for the occurrence of a cross-talk between adaptive immunity and the intestinal microbiota.[200] This hypothesis is strengthened by the observation that the gut microbiome appears to influence the adaptive immune system, since intraepithelial $CD4CD8\alpha\alpha^{+}$ double positive T lymphocytes required the presence of *L. reuteri*, in combination with a tryptophan-rich diet, to reprogram intraepithelial CD4+ T cells into immunoregulatory T cells.[201] A link between behavioral responses, the microbiota, and the adaptive immune system has been illustrated in a study showing that chronic defeat stress in mice may transiently elevate $IL\text{-}10^{+}$ T regulatory cell levels, associated with a lower abundance of Clostridial species, reinforcing the hypothesis that the gut microbiota may interact with the adaptive immune system to influence brain functions.[202] B and T cell-deficient $Rag1^{-/-}$ mice subjected to a water avoidance stress test displayed altered baseline behaviors, including memory and anxiety, overactive HPA axis, increased intestinal secretory state, dysbiosis, and decreased hippocampal c-Fos expression.[203] All these peripheral and central dysfunctions were normalized by pretreatment with probiotics, *L. rhamnosus and L. helveticus*, indicating a role for adaptive immune cells in maintaining normal intestinal and brain health and that probiotics can overcome the immune-mediated deficit along the gut-brain-microbiota axis.[203] Recently, in a murine model of inflammatory bowel disease (IBD), adaptive immune responses to specific bacterial antigens could be modified by appropriate dietary components, suggesting that dietary manipulations could be therapeutically valuable for human diseases.[204] Overall these studies strengthen the importance of focusing on the role of the microbiome in the fine tuning of the immune system modulation of both intestinal and brain functions.

3.4 Microbiota influence on the enteric nervous system

The enteric nervous system (ENS) is a complex network constituted of ganglia and interconnecting fiber strands which may respond directly, or indirectly, to several metabolites originating from the gut microbiota.[1] In the gut, the ENS innervates smooth muscle and epithelial cells, intrinsic blood vessels and EECs and is deputed to the control of different functions such as motility, gastric secretion, transport of fluids across the epithelium, blood flow, nutrient absorption, in a rather autonomous way with respect to the CNS.[205] Gastrointestinal reflexes, such as the peristaltic reflex, or generation and progression of migrating myoelectric complexes (MMC) during interdigestive fasting periods depend upon the intrinsic innervation.[206] However, the ENS is not totally autonomous and the complete neuronal control of gut functions originates from the integration of local reflexes, with reflexes mediated by intestinofugal neurons, sending inputs to sympathetic ganglia, and by vagal, splanchnic and pelvic nerves afferent sensory pathways to the CNS.[206,207] The human ENS is composed of 200–600 million neurons, the same number present in the spinal cord. Three major plexuses: the subserous, the myenteric (located between the longitudinal and circular smooth muscle layers) and the submucosal (located in the homonymous layer) concur to the formation of the ENS.[205] Enteric neurons are classified according to their morphology, neurochemical coding, projections to targets and functional roles, into at least 20 distinct types of neurons. Neurons in the ENS have been distinct in four major functional types: intrinsic primary afferent neurons, interneurons, excitatory and inhibitory motor neurons.[206] Intrinsic primary afferents are sensory neurons, which detect diverse stimuli (i.e., chemical and mechanical) in both the mucosa and muscularis propria and initiate appropriate motor, secretory and vasomotor reflex responses.[208] Importantly, as more recently proposed, the ability of intrinsic neural pathways together with extrinsic primary afferents, to sense factors derived from the gut lumen, including microbial metabolites, does not only influence to gut function but also the CNS.[6] The chemical coding of primary afferent enteric neurons is highly conserved in different animal species, and is represented by cholinergic and serotoninergic neurons, peptidergic neurons containing tachykinins and calcitonin related gene peptide (CGRP).[208–210] Excitatory and inhibitory motor neurons receive fast excitatory synaptic potentials and innervate the longitudinal and circular smooth muscle layers and the muscularis mucosae along the whole gastrointestinal tract.[206] In motor neurons, primary excitatory neurotransmitters are ACh and tachykinins, while several neurotransmitters have been identified in inhibitory neurons, including nitric oxide (NO), vasoactive intestinal peptide (VIP) and ATP-like transmitters, although NO is considered the primary transmitter.[205,211,212] Local reflexes are coordinated by several types of ascending and descending interneurons, with different chemical coding, according to their projection and function.[205] A distinctive feature of the ENS is that enteric neurons communicate with different cell types within the enteric microenvironment.[213–215] In this regard, an increasing body of evidence suggests the existence of a cross-talk between commensal bacteria and enteric neuronal circuitries.[97,216] The ENS, for example, represents a fundamental control system for the intestinal microbiota by modulating intestinal motility thus allowing the removal of exuberant bacteria from the lumen.[217] More recently, the ENS has been suggested to influence the microbiota composition.[218] In zebrafish carrying a genetic mutation which prevents the formation of the ENS ($sox10^{-/-}$) the microbiota showed high level of gut inflammation associated with a "pro-inflammatory" microbial lineages. The inflammatory phenotype could be attenuated by transplantation of wild-type ENS precursors or by the administration of "anti-inflammatory" bacterial strains.[218] More information is, however, available on the ability of the enteric microbiota to influence the development and function of both sensory and motor gut functions. In particular, the saprophytic microflora seems to modulate the development of the ENS in critical postnatal life times, in the first 3–4 years of life, when the density of enteric neurons may significantly decrease.[211,219] This hypothesis is substantiated by the evidence that newborn GF mice display important anomalies in the ENS, with a decreased nerve density and neuron number per ganglion, changes in the neurochemical coding, alterations in intrinsic primary afferents and in the contractile responses, which are not observed in specific pathogen free (SPF) controls and after conventionalization of GF animals.[220,221] Complex morpho-functional neuromuscular rearrangements of the ENS have been demonstrated also after chronic antibiotic treatment inducing dysbiosis in juvenile mice.[222] In this study, dysbiosis was associated with distortion of the enteric glial network, reduction of myenteric neurons, reduced number of $nNOS^{+}$ neurons, altered cholinergic, tachykininergic and nitrergic neurotransmission. More recently, in adolescent mice dysbiosis was associated with up-regulation of the neurotrophic pathways brain-derived neurotrophic factor (BDNF) and its high-affinity receptor tropomyosin-related kinase B (TrkB) within specific substance P and CGRP-containing myenteric neurons, which are involved in motor and sensory gut functions.[216] Dysbiosis-induced neurochemical derangement in the ENS was associated with changes in ileal TLR2 receptor distribution and slowing of the gastrointestinal transit, which was partially recovered by activation of TLR2 signaling.[222] Indeed, the ability of specific bacterial species to influence the ENS development seems to principally involve activation of PAMP receptors, such as TLR2 and TLR4.[219] Several studies resorting to TLR mutant mice are available to suggest the importance of TLRs in maintaining ENS homeostasis.[223] GF, antibiotic-treated mice and TLR4$^{-/-}$ mice showed a similar pattern of small intestine and colonic motor changes with alterations in number of nNOS^{+} neurons and gliosis.[224,225] Derangement of ENS structural and functional integrity, similar to that observed in antibiotic-induced microbiota depleted wild-type mice was also observed in TLR2$^{-/-}$ mice.[226] Interestingly, treatment with low LPS levels favored the survival of primary enteric neuron cultures with an NF-kB–dependent mechanism.[224] Besides TLRs, metabolic products of intestinal bacteria, such as SCFAs or chemotactic peptides (for example, N-formylmethionyl–leucine–phenylalanine) are able to stimulate the ENS and influence the efficiency of the gastrointestinal transit.[2,227] Specific bacterial strains may be involved in maintaining the normal intestinal motor function. In GF rats, the derangement of the neuromuscular function leading to a significant delay in the intestinal transit and MMC period, were partially reversed after colonization with either *Lactobacillus acidophilus* or *Bifidobacterium bifidum*. On reverse colonization with *Escherichia coli* and *Micrococcus luteus* delayed gut motility.[228] The function of intrinsic primary afferent neurons may also be influenced by the gut microflora. In GF mice the electrophysiological properties of myenteric primary afferent neurons in vitro displayed reduced excitability that was restored after colonization with normal gut microbiota.[229] *Lactobacillus reuteri* increased the excitability and the number of action potentials per depolarizing pulse, decreased calcium and potassium channel opening, and reduced slow afterhyperpolarization in intrinsic primary afferents neurons.[230] In an electrophysiological study, *Bacteroides fragilis* activated intrinsic primary afferents neurons in vitro and this effect was exclusively dependent upon polysaccharide A, a bacterial component, and did not depend on the diffusion of molecules from the epithelium to the ENS.[231] This suggests that epithelial elements may represent a fundamental interface for the transduction of microbial signaling from the lumen to the ENS as demonstrated in a further study evaluating the effect of the L. *rhamnosus strain* (JB-1) on ENS, where the in vivo and in vitro effect of oral microvesicles derived from JB-1 confirmed that an intact epithelium was required to mediate the effect of the probiotic.[232] However, there are also reports suggesting that metabolites produced by the gut microbiota may influence intrinsic primary neurons in the ENS. Patch clamp studies in rat mucosa hemidissected preparations showed that epithelial application of the microbial-derived SCFA, butyrate, as well as 5-HT, could affect myenteric intrinsic primary afferent excitability providing some further insight into mechanisms by which the microbiota may interact with the ENS.[230]

4. Connections along the microbiota-gut-brain axis

A complex reflex neural network contributes to the formation of the gut-brain axis, allowing a two-way communication system between the gut and the brain, which are in constant cross-talking in both health and disease states[2,233] (Fig. 2). This bidirectional communication axis allows sensory visceral signals from the gut to influence the brain in order to regulate reflex activity and mood states, and, in turn, inputs from the brain may be directed to the periphery to modulate gastrointestinal motor, secretory, sensitive and immune functions.[2,205] Afferent and efferent neurons, within the parasympathetic (vagal) and sympathetic (splanchnic and pelvic spinal pathways) branches of the autonomic nervous system (ANS) convey the neuronal signaling. In the CNS, both spinal and vagal afferent inputs synapse with higher brain regions, in particular with the emotional motor system, consisting in the limbic system and in some paralimbic structures (including the medial prefrontal cortex, amygdala, and hypothalamus), which coordinate responses to emotion along the gut-brain axis.[234,235] In particular, afferent vagal neurons have their cell bodies in the nodose vagal ganglion (NVG) and transmit sensory information to the nucleus of the solitary tract (NTS) in the brain stem regarding the presence of food, motor activity and degree of gut distension. This information is then conveyed to higher centers, such as the paraventricular nucleus of the hypothalamus areas, representing the main source of corticotrophin releasing factor (CRF) or, more locally, to form long vago-vagal reflexes.[205,234,235] Efferent vagal pathways drive signals to the periphery, via the dorsal motor nucleus of the vagus (DMV) and the nucleus ambiguous of the brain stem, underlaying control of motor and secretory gut functions.[205,234,235] Afferent spinal neurons have their cell bodies in the dorsal root ganglia (DRG) and contribute to transmit sensory inputs to the dorsal horn neurons of the thoracic and upper lumbar spinal cord, which then project with second-order neurons to the brain via the spinoreticular, spinomesencephalic, spinohypothalamic, and spinothalamic pathways.[1]

The first three of these ascending tracts project to the brainstem in the dorsal reticular nucleus, to the periaqueductal gray and other midbrain regions, and to the hypothalamus, and modulate unconscious and/or automatic responses to visceral sensory inputs, i.e., arousal, orientation, autonomic responses, prototype behavioral, and emotional responses, including the emotional component of pain.[1,236,237] The spinothalamic tract terminates in the medial and posterior thalamus, from which the thalamocortical fibers project to the primary somatosensory cortex. The thalamus is the relay station where multiple somatic and visceral inputs, including nociception, converge before reaching the cortical regions such as, the anterior cingulate cortex, the insula, and somatosensory cortex where painful stimuli are elaborated.[237]

Another fundamental pathway contributing to the cross talk between the brain and the gut is represented by the hormonal hypothalamic–pituitary adrenal (HPA) axis. Corticotrophin releasing factor (CRF) release from the paraventricular nucleus (PVN) of the hypothalamus primes the activation of the HPA axis. The consequent release of adrenocorticotrophic hormone (ACTH) from the anterior pituitary and its diffusion into the systemic circulation leads to the release of glucocorticoids from the adrenal cortex. This hormonal cascade has an important role in the regulation of several functions of the brain–gut axis, controlling behavioral stress responses, but also related changes in the gastrointestinal transit, visceral sensation and permeability of the intestinal wall.[2,235]

Numerous studies have highlighted the fundamental influence that the enteric microbiota exerts on both the ANS and HPA axis function in the context of the microbiota-gut-brain bidirectional communication.[6] In the following paragraphs the major evidences supporting the bacteria mediated regulation of vagal pathways, and HPA axis will be discussed.

4.1 The vagus nerve

The majority of studies concerning the bi-directional influence exerted by the ANS on gut and brain functions have focused on the role of the vagus nerve, which, owing to its extensive innervation, is responsible for the collection of information from several peripheral organs, including the respiratory, cardiovascular and gastrointestinal system.[238] The hepatic and celiac branches of the vagus nerve innervate the entire gut, except the transverse and distal colon, and underlay control of motor and secretory gut functions.[205] Vagal afferents mainly regulate feeding behavior by modulating upper gut reflexes, e.g., gastric accommodation, gastric emptying, gastric/pancreatic secretion, emesis. Three types of vagal fibers innervate the mouse gut: mechanoreceptors, tension receptors and mucosal varicose nerve endings.[205] Vagal mechanical afferents terminate within the muscle layers and comprise intramuscular arrays and intraganglionic laminar endings. Intramuscular arrays innervate the longitudinal or circular muscle layers and respond to muscle stretch. Intraganglionic laminar endings terminate on the basal lamina surrounding myenteric ganglia and are sensitive to distension and muscle contraction. The tension-sensitive afferents, which are principally sensitive to physiological distension, are activated by normal peristaltic activity.[239,240] Vagal mucosal afferents have been detected in the stomach mucosa and in the villus and crypts of the small intestine mucosa, and serve to detect chemical stimuli, such as gastric acid, to regulate satiety, to allow food particle passage into the duodenum and to sense local hormones secreted by the mucosa, luminal molecules such as bacterial metabolites, or neuromodulators/neurotransmitters, to start vagal reflexes.[205] However, afferent nerves are not exposed to the gut luminal side, thus, in normal circumstances sensory neurons are indirectly activated by stimuli in the intestinal lumen in a paracrine modality, which is mediated by compounds such as 5-HT, cholecystokinin, histamine, somatostatin and CRF, all of which can be released from neuroendocrine cells in the mucosa.[241,242] ECs signaling to vagal afferents potentially provides a direct pathway that connects chemical stimuli in the intestinal lumen with the supraspinal networks involved in reflexes, such as vomiting.[2] Indeed, nerve terminals of vagal afferents are located in close proximity to ECs cells, and these terminals express 5-HT_3 receptors, which represent the target for antiemetic drugs.[2,243] Recently, vagal endings have been described to form connections with a subset of ECCs, denominated neuropods, which provide a neuroepithelial circuit for fast sensory transduction mediated by cholecystokinin (CCK) and peptide YY (PYY).[128] Although pain evoked by distension of the upper gastrointestinal tract is probably mediated via sympathetic splanchnic afferents, vagal afferents may participate to pain reactions evoked by chemical stimuli, such as gastric acid challenge.[244] In addition, activation of lower gut vagal afferents has been suggested to modulate visceral pain spinal transmission, since, after subdiaphragmatic vagotomy in rats, pain responses to colorectal distension increased.[207]

It is now ascertained that the vagus nerve retains a central role along the microbiota-gut-brain axis, since signals originating from the microbiota may reach the CNS via the vagal pathways and vice versa.[6] Indeed, the vagus nerve may participate to microbiota-induced behavioral changes. After oral administration of *Campylobacter jejuni* to mice, changes in *c-Fos* expression were observed in the cell bodies of vagal afferents and in the NTS contributing to an early signal to the brain in the presence of potential pathogens.[245] In a previous study, mice receiving an oral administration of *C. jejuni* displayed an anxiety-related behavior, in the absence of an overt immunological response.[246] In a more recent study, rats subjected to subdiaphragmatic vagal deafferentation displayed altered gene expression in the nucleus accumbens, contributing to brain and behavioral anomalies.[247] The microbe-driven effect on the vagus may favor a rapid communication of signals to the brain, unlike immune and endocrine pathways, and may underlay the positive effects of probiotics on the CNS.[248] For instance, in mice, administration of *Lactobacillus rhamnosus* (JB-1) increased the firing rate of the mesenteric nerve bundle, which contains vagal and spinal afferents[249] and determined neurochemical and behavioral effects, which were dependent on intact vagal nerve activation.[250] In transgenic mice Shank3B$^{-/-}$, exhibiting autism spectrum disorders (ASD)-like behavior, administration of L. *reuteri* rescued social interaction inducing synaptic plasticity in the ventral tegmental area in a vagus-dependent manner.[251] Vagus integrity was also required by the prebiotic 2-fucosyllactose, to promote LTP and associative learning skills in rats.[252] However, oral administration of antimicrobials to innately anxious male BALB/c mice altered the microbiota composition and increased the exploratory behavior and hippocampal BDNF levels, independently from vagal integrity suggesting that the vagus may only partially modulate bacterial effects in the along the microbiota-gut-brain axis.[253]

In GF mice the excitability of vagal afferent neurons that synapse with IPANs in the ENS was significantly reduced with respect to the SPF and conventionalized ex-GF animals suggesting that an intact microbiota is crucial for normal intrinsic and extrinsic nerve function.[221] Since these sensory neuronal pathways are essential in the control of gut functions, such as motility and secretion, it is feasible to hypothesize that any change involving the enteric microbiota composition and function may, also, predispose the host to altered gut motility and secretion, which results in diarrhea or constipation. These changes are, in turn, likely to influence the balance of the enteric microbiota.[2] Impaired intestinal transit caused by compromised MMC motor complexes, which are controlled by vagal pathways, is associated with bacterial overgrowth in the small intestine.[254] Such alterations in the gut motor function may also influence the delivery of pre-and probiotics to the small intestine and colon, including resistant starches and dietary fibers, and other microbial nutrients which have a functional impact on the gut microbiota.[6,255]

4.2 Hormonal connections: The HPA axis

A role of intestinal microbes in controlling the development and function of the HPA axis has been demonstrated in different experimental conditions, mainly resorting to GF animals.[256] Indeed, GF mice exhibited an increased HPA axis activity, with elevated levels of plasma corticosterone, in response to stressful stimuli.[256,257] Stress-related changes of the microbiota-gut-brain axis may have important consequences on the expression and activity of glutamatergic NMDA receptors and the closely related neurotrophin, brain-derived neurotrophic factor (BDNF), both fundamental for neuroplasticity in the CNS.[258–260] In the hippocampus of young-adult BDNF heterozygous mice, corticosterone-induced postnatal stress was associated with downregulation of BDNF expression and changes in NMDA receptor subunit expression.[259] IBS patients, who commonly exhibit high levels of anxiety and depression, often related to high sensitivity to stressor stimuli;[261,262] as well as alterations in the microbiota composition,[263] exhibited higher ACTH and cortisol levels after CRF infusion with respect to healthy subjects.[264] The immune system may also have a role in the interplay between the gut microbiota and the HPA axis occurring as a consequence of stress stimuli. In this condition, enhancement of the permeability of the epithelial barrier allows bacterial translocation and the consequent activation of mucosal immune responses may activate the HPA axis.[6] Activation of the HPA axis as well as the release of ACTH and glucocorticoids are also under the control of the vagus nerve, suggesting the existence of a cross talk between neuronal and hormonal pathways modulating the gut-brain communication.[265]

5. Neuroactive molecules along the MGB axis

Increasing evidence suggests that the gut microflora may contribute to the production of neuroactive molecules as a result of evolutionary adaptations, including late horizontal gene transfer.[266] The ability of microbes to produce and respond to neuroactive and neuroendocrine molecules, including noradrenaline and adrenaline, proposed about 30 years ago, formed the basis for the existence of a gut-brain bidirectional communication system.[267,268] Since then several other neuroactive molecules, among others, a variety of amines such as 5-hydroxytryptamine (serotonin, 5-HT) and kynurenine, deriving from tryptophan metabolism, neurotransmitters, such as glutamate, GABA and acetylcholine, neuroactive short chain fatty acids (SCFAs), secondary bile acids, have been shown to be produced by the gut microflora.[269] These microbially-derived molecules may influence brain activity giving rise to a microbiota-mediated bottom-up control of the CNS.[269] Conversely, several signaling molecules used by the host for neuronal and neuroendocrine signaling (for example, catecholamines, 5-HT, dynorphin, and cytokines), which are secreted into the gut lumen by neurons, immune cells and enterochromaffin cells, may directly interact with certain luminal microbes via intercellular signaling mechanisms.[66] The next paragraphs give an overview of the role of some of the most important neuroactive molecules and relevant targets in the modulation of this interkingdom cross-talk.

5.1 Tryptophan, serotonin and kynurenine

L-tryptophan, is an essential amino acid which cannot be synthesized by the human body and is provided via nutritional sources from the diet.[270] Although L-tryptophan concentrations in human tissues are lower than those of other amino acids, the presence of this essential molecule is fundamental for human health. L-tryptophan plays a limiting role in the synthesis of proteins but it is also a precursor for the production of important bioactive molecules influencing gut-brain communication.[271] Tryptophan can be readily metabolized within the gut along different metabolizing pathways, two of the majors being represented by the kynurenine[269] and serotonin (5-HT) biosynthetic pathways.[272] Most of the absorbed tryptophan (about 90%) is metabolized along the kynurenine biosynthetic pathways upon activation of indoleamine-2,3-diooxygenase (IDO1) and tryptophan-2, 3-dioxygenase (TDO),[273] while only approximately 3% is metabolized into serotonin (5-HT) via tryptophan hydroxylase 1 (TPH1), the rate-limiting enzyme in the synthesis of 5-HT, and the rest is degraded by the gut microbiota to produce indole and its derivatives.[274]

5.1.1 Serotonin

In humans 5-HT is mainly concentrated in the GI tract, where it is released from ECs cells of the mucosa and by myenteric neurons, playing a key role in the modulation of several gut functions, including sensory, secretory and motor function.[159,272] The fraction of free tryptophan reaching the circulation may cross the BBB to participate to 5-HT synthesis in the brain.[271]

In the CNS, the biogenic amine is involved in the modulation of mood, behavior and cognitive functions and alterations of the serotoninergic transmission participate in the response to stressful stimuli as well as in the pathogenesis of psychiatric disorders, including major depression and schizophrenia.[13]

An increasing number of studies suggest that the gut microbiota participates to the regulation of different mechanism influencing 5-HT concentration in the gut. Indigenous spore forming clostridia-taxa from the mouse microbiota favored 5-HT biosynthesis from colonic ECs and gastrointestinal motility, via the production of deoxycholate.[275] This result reflects data obtained by other authors demonstrating that gut bacteria may produce secondary bile acids influencing 5-HT synthesis in the host, which involves a TGR5-mediated pathway.[276] The ability of metabolic products released by the microbiota to influence 5HT gut levels has also been shown in a model of ex-GF mice colonized with human gut microbiota. In this condition colonic TPH1 mRNA and protein levels as well as 5-HT intraluminal concentrations increased with respect to the corresponding GF animals. In the same study, enhancement of TPH1 mRNA was reproduced by exposing a human EC cell line to SCFAs.[277] Interestingly, the release of 5-HT from human ECs isolated from Crohn's disease patients was stimulated in vitro by LPS via TLR4 activation.[278] The gut microbiota may also influence 5-HT metabolic pathways. 5-HT luminal levels in the cecum and colon were significantly lower in GF mice than in gnotobiotic mice recolonized with specific pathogen free (SPF) or with fecal flora (ex-GF).[279] In particular, recolonization resulted in a major fraction of free unconjugated biologically active 5-HT, suggesting that bacteria may contribute to maintain free 5-HT in the gut lumen, by promoting the deconjugation of glucuronide-conjugated 5-HT.[279] The microbiota influences 5-HT levels in the gut also by regulating the biogenic amine transporter system in intestinal enterocytes as shown in a post infectious IBS rat model after administration of a probiotic strain of Lactobacillus.[280] The gut microflora may influence the function of the serotoninergic transmission in the ENS. In a recent study, colonization of GF mice with a microbiota derived from conventionally raised mice modified the neuroanatomy of the ENS and increased motor activity, which was dependent on both mucosal and neuronal release of 5-HT and the proliferation of Nestin$^+$ enteric neuron progenitors in the adult gut.[281] This latter observation confirms the well-established concept that 5-HT has a prominent role in neurogenesis and neuroprotection in the adult ENS.[13,149,281] It remains, however, to elucidate if changes of 5-HT production in the gut may have a direct effect on the brain, because 5-HT cannot cross the BBB.[282] Changes in peripheral levels of 5-HT may, however, influence gut-brain communication. For example, decreased 5-HT gut levels have been detected in a mouse model of ASD (BTBR T^+ *Itpr3*^{tf}/J mouse) which significantly correlated with the relative abundance of bile-metabolizing Bifidobacterium and Blautia.[283] In addition, the relevance of peripheral 5-HT_3 ionotropic receptors in the modulation of vagal afferents, to modulate chemotherapy-induced emesis, as well as the ability of peripheral 5-HT_4 and $5HT_3$ receptors to modulate visceral pain associated with functional gut disorders, such as IBS, are quite well established.[13,149,284] In IBS patients, 5-HT homeostasis is highly altered, with lower mucosal 5-HT content, TPH1, and serotonin reuptake transporter expression levels.[285–287] Administration of 5-HT_3 receptor antagonists could slow colonic transit, enhance small intestinal absorption but also reduce visceral pain, implying activation of gut-brain pathways.[288] Since the microbiota participates in pain signaling from the gut, it is tempting to suggest that 5-HT producing microbes, such as Candida, Streptococcus, Escherichia and Enterococcus, may influence visceral pain perception.[13] However, also in view to the uncertain and controversial role of 5-HT signaling in the gut, further research is necessary to confirm this issue.[289] Another aspect of 5-HT signaling along the microbiota-gut-brain axis, is that 5-HT produced by EC cells into the gut lumen may influence the microbiota, although the physiological consequences of this pathway have not yet been clear-cut described[149,272] Some of the few studies available show that 5-HT may modulate bacterial motility and promote the expression of virulence genes in bacteria via a quorum-sensing mechanism.[22,290]

5.1.2 Kynurenine

Tryptophan represents the biological precursor of the kynurenine and indole derivative pathways, under both direct and indirect microbiota control.[291] The biosynthesis of kynurenine is strictly dependent upon activation of specific and tissue-dependent enzymes: IDO-1 being found in most tissues including the intestine, and tryptophan-2,3-dioxygenase (TDO) being located mostly in the liver and responsible for the majority (almost 99%) of kynurenine production.[269] The two enzymes are also distinguishable from a functional view point, IDO represents the rate limiting step of kynurenine formation and is induced by proinflammatory cytokines, while TDO is regulated by glucocorticoids.[13,292] The kynurenine arm of the tryptophan metabolic pathway has important physiologic functions both in the periphery and in the CNS. Indeed, dysregulation of kynurenine metabolism has been suggested to underlay development of gut-brain axis disorders.[269,270,291] Among the downstream by-products of kynurenine, kynurenic acid (KynA) and quinolinic acid are of particular pathophysiologic importance. In normal conditions, KynA and quinolinic acid do not cross the BBB, in considerable amounts.[293] The BBB is, however, permeable to kynurenine, which, principally in astrocytes, is transformed into KynA, and in the microglia is metabolized into quinolinic acid. Quinolinic acid is an agonist at NMDA glutamate receptor, and is considered as neurotoxic proinflammatory molecule.[294] KynA is a natural competitive antagonist, with high affinity for the glycine site associated with NMDA receptor.[295] In addition, KynA acts as an antagonist at α7nACh and to GPR35 receptors.[270] KynA protects neurons against glutamate-mediated excitotoxic damage, and is considered neuroprotective both in the CNS and in the ENS.[215,296,297] Since the microbiota may influence both the integrity of the BBB as well as the maturation and activation states of microglia and astrocytes, an intriguing task, which is still largely unexplored, is represented by the evaluation of the role played by the saprophytic microflora in the modulation of kynurenine pathway homeostasis within the periphery and in the CNS. This point is of fundamental translational value in view of the importance of this pathway throughout life, especially during the early neurodevelopmental phase.[8] Indeed, in rats, prenatal kynurenine has been shown to induce age-dependent changes in the expression of GluN1 and GluN2A subunits of NMDA receptors. Furthermore, prenatal inhibition of the kynurenine pathway results in altered synaptic transmission and brain protein expression and cognitive alterations in the adult offspring.[298–300] The microbiota affects host kynurenine production by influencing both the glucocorticoid-based activity of TDO,[301] and the immune-dependent activity of IDO activation.[302] In GF animals and in dysbiotic animals, circulating tryptophan levels increased leading to a reduction of the kynurenine-to-tryptophan ratio, which has been attributed to changes in IDO and TDO activities.[303–306] Interestingly, administration of probiotics such as *Bifidobacterium infantis* could affect tryptophan metabolism, resulting in normalized tryptophan and kynurenine levels.[306] It is to be established, however, if changes in circulating kynurenine, associated with changes in the microbiota composition and function, may impact on CNS kynurenine homeostasis. It has recently been demonstrated that the gut microbiota may regulate kynurenine pathway genes in the hippocampus through a miRNA dependent pathway.[307] In this study, expression of miRNAs, such as miR-294–5p, Brd2, and Slit3kr, regulating the expression of central kynurenine pathway genes, increased in the hippocampus of GF mice and could be normalized by microbial colonization.[307] There are reports suggesting the metabolic products of the gut microbiota may also regulate the kynurenine pathway. For example, circulating bacterial metabolites, such as SCFAs, can exert an inhibitory effect on IDO and the kynurenine pathway exerted in a STAT1/HDAC-dependent manner.[308] *Lactobacillus johnsonii* was demonstrated to reduce kynurenine circulating levels in rats, as well as IDO activity in HT-29 intestinal epithelial cells, possibly by increasing hydrogen peroxide production.[309,310] Emerging evidence suggests that the diversion of the tryptophan metabolism from the 5-HT pathway towards the kynurenine pathway may have an important role in the manifestation of psychiatric disorders such as anxiety and major depression.[291] In chronically stressed mice, displaying despair behavior, alterations in the microbiota composition and metabolic signature (i.e., reduced Lactobacillus and increased circulating kynurenine levels) were restored after supplementation with *L. reuteri,* which also ameliorated behavioral abnormalities, suggesting that manipulation of the microbiota may contribute to regulating metabolism and resilience during stress.[311] In GF mice, induction of depressive mood after fecal transplantation was associated with an increase in the kynurenine/tryptophan ratio.[233] Accordingly, in vitro and in vivo studies have demonstrated that activation of IDO1 in the presence of IFNγ facilitates the conversion of tryptophan into kynurenine, promoting the development of depressive symptoms.[312] Furthermore, increased anxiety-like behaviors developed in mice receiving gut microbiota from depressed patients in parallel with increased kynurenine levels and an increased kynurenine/tryptophan ratio in blood circulation.[313] Administration of *Bifidobacterium infantis* ameliorated the depressive behavior in rats subjected to forced-swim test, which was correlated with normalization of tryptophan metabolism.[306]

5.2 Catecholamines

Cathecolamines participate in the regulation of host physiology, including regulation of motor activity, cognitive and executive functions, attention-related processes and stress responses.[314,315] Catecholaminergic pathways are also involved in the modulation of several key functions in the gut, regulating the vascular tone, inhibiting the motor function and fluid secretion.[316] Enteric neurons are endowed with the whole neurotransmitter machinery for noradrenaline and dopamine, but lack the *N*-phenylethanolamine transferase, which is fundamental for the synthesis from noradrenaline of adrenaline.[317] The catecholaminergic system is abundantly expressed in the gut and was one of the first neurotransmitter pathway, which was shown to participate to the crosstalk between the microbiota and human gut.[60] In the course of gut homeostasis alterations, the release of noradrenaline into the gut lumen, has been suggested to promote pathophysiological responses in the host, by influencing the composition and stability of bacteria, including the promotion of a pathogenetic growth.[66,318] For example, noradrenaline, released in the gut lumen consequently to a surgical trauma, induced the expression of virulent traits in *Pseudomonas aeruginosa*, resulting in gut derived sepsis.[217] In addition, noradrenaline increased the expression of Shiga toxins produced by *E. coli* O157:H7 (Lyte et al. 1996), and of several pathogenic Vibrio species (Nakano et al. 2007). In vitro noradrenaline favored the growth of several strains of enteric pathogens and enhanced the virulence of *Campylobacter jejuni*.[319] The mechanisms underlying catecholamine induced expression of virulence traits in bacteria are diverse and include the ability to act as a quorum sensing molecule. Both adrenaline and noradrenaline were necessary for *C. rodentium* to fully activate its virulence program and successfully colonize its murine host, in a QseC/QseE-dependent manner.[60] Catecholamines may also promote bacterial growth by acting as enterobactin-like pseudosiderophore and inducing iron release from transferrin, which can then be taken up by bacteria by either ferric or ferrous iron uptake systems.[320,321] This bacterial iron theft may induce infectious disease processes, since catecholamine-released iron enables fewer than 100 bacteria in 24 h to increase their cell numbers by more than 105-fold.[322] Finally, catecholamines may induce bacterial migration towards the host intestinal mucosa exerting a chemoattractant effect.[323] In this latter regard, noradrenaline was found to increase the intestinal mucosa adherence and enteropathogenicity of *E. coli* O157:H7, whereas Green et al.[324] and Chen et al.[325] demonstrated that catecholamines increase the adhesion and invasiveness of *E. coli* and Salmonella to the mammalian gut.

The relevance of the catecholaminergic system as an interchangeable molecular pathway in the crosstalk between the microbiota and human gut is shown also by the demonstration that the enteric microbiota may produce catecholamines with chemical structures similar to those produced by the host.[60] A strain of Escherichia K-12 was shown to synthesize noradrenaline, dopamine[326] and DOPA, the biological precursor of catecholamines, suggesting that the biosynthesis of monoamine neuromediators in bacteria involves enzymatic systems analogous to those present in animals. In another study, Bacillus was shown to produce both noradrenaline and dopamine.[327] In GF mice the levels of free noradrenaline in the gut lumen were lower than in their specific pathogen free controls, further strengthening the concept that bacteria possess catecholamine synthesizing enzymes.[328] Bacteria also express plasma membrane monoamine transporter and organic cation transporter proteins, which clear noradrenaline from the extracellular space, thus enabling the catecholamine transfer into microbes.[329] In addition, bacteria may promote the transition of the catecholamine structure from an inactive to an active form. In GF mice, luminal catecholamines were biologically inactive and could be converted into the biological active form after exposure to β-glucuronidases produced by either *Clostridium species* or SPF fecal flora.[328] Although catecholamines play important role in the regulation of gut physiology functions,[316] the influence that bacterial derived catecholamines may have on the host gut function is largely to be uncovered. Intraluminal administration of dopamine increased colonic water absorption in an in vivo ligated loop model of SPF mouse model, thus suggesting that luminal catecholamine plays a role as a pro-absorptive modulator of water transport in the colon.[328]

In addition to the possible involvement of catecholamine in the local interplay between the gut host and the microbiota, several reports are available to show that changes in the gut microbiota homeostasis may alter catecholamine levels in the brain. In both GF free and dysbiotic animals, development of alterations in the behavioral phenotype were associated with changes in brain neurochemistry, involving alterations of catecholamine levels. In rats, long term treatment with antibiotics induced cognitive deficits and anxiolytic effect which were associated with changes in CNS neurochemical patterns, including enhanced noradrenaline levels in the hippocampus and of L-DOPA in the amygdala.[330] Analogously, GF mice displayed increased motor activity and a reduced anxiety-like behavior compared to their SPF controls, which were associated with an increased turnover of noradrenaline, dopamine and 5-HT in the striatum, reduced BDNF levels and alterations in the expression of genes related to synaptic plasticity, such as synaptophysin and PSD-95, in different brain regions related to anxiety.[331] In a more recent study administration of *Lactobacillus helveticus R0052* and *Bifidobacterium longum R0175* attenuated the effects of water avoidance stress. These beneficial effects were correlated with a reduction in the plasma levels of stress neuroendocrine markers such as, corticosterone adrenaline and noradrenaline.[114] Overall, these observations further lend support to the existence of a close connection between behavior and the microbiota, where catecholamine pathways may represent a possible bridging molecular system.

5.3 GABA and glutamate

GABA is the most important inhibitory neurotransmitter in the brain, and dysfunction of GABAergic neurotransmitter pathways has been implicated in pathogenesis of several brain disorders, such as epilepsy and schizophrenia.[332] Both prokaryotes and eukarya synthesize GABA through the decarboxylation of glutamate (Glu) by glutamate decarboxylase (GAD). GAD has been found in both Gram-positive and Gram-negative bacteria, where it is associated with systems that are involved in pH homeostasis and the generation of metabolic energy.[333,334] Interestingly, luminal GABA levels are significantly reduced in GF mice compared to colonized ex-GF mice.[335] *E. coli*,[336] Pseudomonas[337] and specific strains that are generally recognized as safe or health-promoting, such as LAB (e.g., strains belonging to Lactobacillus, Lactococcus, and Streptococcus genera) and Bifidobacterium can synthesize GABA[338–340]. Accordingly, a metagenomic analysis suggests that a high proportion of human gut microbiota possesses genes encoding for GAD.[341] In particular, one Lactobacillus strain and four strains of Bifidobacterium, isolated from the human intestine, were described to produce GABA.[338] Fermented foods of the traditional Chinese cuisine, which are substrate for Lactobacillus spp., contain high millimolar levels of GABA.[342] Although the importance of GABA production by bacteria in the host intestinal lumen is unclear, modulation of GABAergic pathways along the microbiota-gut-brain axis may influence the host health.[340] In rats, the oral administration of *B. dentium*, a bacterial species isolated from human feces and able to synthesize GABA, inhibited visceral hypersensitivity, suggesting that targeting GABAergic signals along this microbiome-gut-brain axis may represent an approach for the treatment of abdominal pain.[341] The ability of orally administered *L. rhamnosus* to increase central GABA levels was demonstrated by magnetic resonance spectroscopy in mice.[343] Bacteria-induced enhancement of GABA levels in the CNS may have functional consequences, as demonstrated by the reduced anxiety and depression related behavior as well as by the increased GABA receptor levels in the hippocampus after oral administration of *L. rhamnosus* (JB-1) in mice.[250] Both the behavioral and molecular effects depended upon the integrity of vagal pathways, since both were abolished after vagotomy.[250] In the juvenile rat hippocampus *L. rhamnosus* and *B. longum* reversed antibiotic-induced reductions in $GABA_A$ receptor α5 and δ subunits.[344] The prebiotics, fructo-oligosaccharide (FOS) and galacto-oligosaccharide (GOS), increased $GABA_{B1}$ and $GABA_{B2}$ receptor gene expression in the hippocampus.[345] These latter findings are all the more interesting since, in the gastrointestinal tract, $GABA_B$ metabotropic receptors are thought to regulate gut motor function and gut-to-brain signaling, including sensitivity of vagal and spinal pathways.[346] Recently, a probiotic *E. coli* strain, Nissle 1917 (EcN), was shown to produce a $GABA_B$ associated analgesic lipopeptide, C12AsnGABA OH, which facilitated GABA diffusion across the epithelial barrier, allowing the activation of GABA receptors on sensory neurons and inhibiting development of visceral hypersensitivity in mice.[347] Besides the neuronal localization, GABA receptors localized in a wide range of immune cells, such as dendritic cells, mast cells and T-cells[348] may be involved in the regulation of immunological processes, such as the down-regulation of pro-inflammatory cytokine release.[349] In this view, GABA is suggested to take part to a neuro-immune cross-talk in the gut.[348] GABA sensing receptors have been also individuated on bacteria,[350] suggesting that host-derived GABA may impact on the microbiota. Interestingly, addition of GABA to *Pseudomonas aeruginosa* cultures increased the microorganism virulence.[351]

Recent attention has also been given to Glu as a neuroactive molecule along the microbiota-gut-brain axis.[1] Diet-derived Glu represents the main source for the aminoacid in the gut, where it undergoes a consistent first pass effect in the splanchnic bed, contributing to the production of energy, necessary for supporting intestinal functions.[352–354] The observation that some bacterial strains of the Lactobacillus spp., identified also in some fermented foods, such as *Lactobacillus plantarum, Lactobacillus paracasei,* and *Lactococcus lactis*[297,355,356] may produce the aminoacid and express Glu-sensing channels, suggest that Glu may represent a signaling molecule underlying the cross-talk between the gut microbiota and the host.[333,357] As a major excitatory neurotransmitter in the CNS, Glu plays a fundamental role in the modulation of both physiological (e.g., memory, learning) and pathologic (e.g., stroke, epilepsy, neurodegenerative diseases etc..) conditions.[358] However, increasing evidence suggest that Glu may also have a role in the regulation of a number of functions in the peripheral nervous system, including the ENS both in health and disease states.[295,296,359,360] Glu acting on both NMDA and metabotropic receptors may be involved in local gastrointestinal reflexes underlying regulation of motor function.[361–363] Recently, in a rat model of postinfectious IBS induced by intragastric administration of *Trichinella spiralis*, development of hypercontractility of the colonic muscularis depended upon synaptic spouting and NMDA receptor potentiation in the ENS downstream to the activation of the receptor tyrosine kinase, EphrinB2/ephB2, involved in structural and functional neuronal plasticity.[364] In addition, several reports, from different groups, suggest that Glu may behave as a sensory co-transmitter within intrinsic primary afferent neurons, transmitting information from the mucosa to the ENS and in extrinsic primary afferents of sympathetic and vagal pathways projecting to higher centers, thus playing a role in the modulation of visceral sensitivity.[365–367] In this scenario, microbes, by modulating glutamatergic pathways along the gut-brain axis, may modify both local functions and behavioral responses.[1] In a rat model of IBS induced by maternal separation, which is associated with alterations of the gut microbiota composition, the levels of the GluA2 subunit of AMPA receptors significantly increased in the hippocampus and their blockade with the antagonist, CNQX, reduced visceral pain perception to colorectal distension, suggesting a possible involvement of ionotropic Glu receptor in central mechanisms of chronic visceral pain control.[368] Besides visceral hypersensitivity, which represents the most common symptom, IBS patients may experience stress-related disorders, with anxiety and depression levels significantly higher with respect to healthy subjects, regardless of IBS-subtype.[262,369] Stress conditions are, indeed, associated with important alterations of central neurotransmitter/neuromodulator pathways, including glutamatergic pathways.[9] A number of studies have demonstrated changes of NMDA, AMPA, mGlu2/3, mGlu5 and mGlu7 receptor expression and function in brain regions such as hippocampus, amygdala and cingulate cortex after stress stimuli.[370–373] Stress-related changes of the microbiota-gut-brain axis may influence the expression and activity of NMDA receptors as well as of brain-derived neurotrophic factor (BDNF), a neurotrophin involved in neuroplasticity, whose function is strictly correlated to NMDA receptor activation in different CNS regions.[258,259,374,375] Corticosterone-induced postnatal stress in young-adult BDNF heterozygous mice was associated with reduced BDNF levels of expression and dysregulation of NMDA receptor expression in the hippocampus.[259] In GF mice, changes in hippocampal BDNF levels were associated with a decreased expression of GluN2A subunit of the NMDA receptor compared to controls.[256] Interestingly, administration of prebiotics, fructo-oligosaccharide (FOS) and galacto-oligosaccharide

(GOS) to rats enhanced BDNF and GluN1 subunit expression levels in the hippocampus gyrus dentate.[376] Analogously, a combination of FOS and GOS displayed a beneficial effect on stress-related behaviors, elevated BDNF mRNA in the mice hippocampus, while administration of B-GOS in rats had pro-cognitive effects, involving upregulation of cortical NMDA receptors.[345,377]

In this view, studies aiming to clarify the functional dynamics and relevance of GABA and Glu and as mediators of host-microbe crosstalk role may eventually lead to the discovery of molecules with a potential therapeutic interest in the treatment of gastrointestinal disease and the related CNS disorders.

5.4 Short chain fatty acids (SCFAs)

It is now evident that the saprophytic microflora in the gut may produce metabolites, which translocating from the intestinal mucosa to the systemic circulation, can act as bacterial "messengers", thus interfering with distant sites including the brain.[269] Among the different gut microbial-derived metabolites, SCFAs, primarily in the form of acetate, propionate, and butyrate, represent the major products generated by bacterial fermentation of dietary fibers in the intestine, as supported by the evidence that GF animals and antibiotic treatment results in drastically lower SCFA levels.[378,379] Due to their metabolism, SCFAs have been implicated in a variety of host processes, such as blood-pressure regulation,[380] circadian rhythm,[381] neuroimmune[382] and gastrointestinal function.[383] Following their production in the colon, in a molar ratio of 60:20:20 for acetate, propionate and butyrate respectively, SCFAs are rapidly absorbed by colonocytes where they can be used by mitochondria as an energy source for the cells.[384] In the gut, SCFAs are described to have a beneficial role, improving gut health by regulating the intestinal barrier integrity, water absorption and mucous production, and gastrointestinal motility. Several in vitro and in vivo studies carried out on rat colon have demonstrated that, SCFAs may modify colonic motor patterns by promoting the release of neuroendocrine modulators, such as GLP-1 and PYY from L cells, and 5-HT from ECs, into the circulation.[385,386]

Besides the local effects on the gut, when transported in the systemic circulation SCFAs can interact with cellular systems in the host, thus influencing gut-brain signaling pathways. SFCAs may directly or indirectly impact on this bi-directional communication axis by regulating different immune, endocrine, epigenetic, and humoral mechanisms.[387] In addition, SCFAs can interact with nerve cells by stimulating the vagus and sympathetic nerve signaling.[35,215] Preclinical studies on GF mice indicate that gut-derived SCFAs may have a role in the maintenance of brain homeostasis, playing a modulatory role in learning and cognition.[388,389] The supplementation of butyrate, acetate and propionate in drinking water can ameliorate reward-seeking behaviors and stress responsivity in mice.[390] In addition to their systemic actions, SCFAs can cross the BBB, although to a minimum extent, passing into CNS and exerting neuroactive properties. Butyrate injection in rat and mouse brains has been shown to enhance mRNA levels of neurotrophic factors, such as BDNF and GDNF, thus improving aspects of neurogenesis, synaptic plasticity, memory formation, and mood-related behaviors, which are generally mediated by these factors.[391] Similarly, in a mouse model of pneumococcal meningitis, butyrate prevented memory impairment reversing hippocampal BDNF and GDNF levels.[392] In the mouse CNS, acetate, propionate and butyrate may also alter levels of neurotransmitters, such as glutamate and GABA,[393] and influence biosynthesis of dopamine, noradrenaline and adrenaline.[394] Alterations of SCFAs have been also associated with a range of CNS disorders, such as Alzheimer's disease[395] or chronic stress,[396] and decreased fecal levels of acetate and butyrate have been also found in the feces of children with ASD.[397] In this latter regard, intraventricular infusions of propionate in mice result in increased repetitive behaviors, impaired sociability, as well as epileptic and convulsive responses, which are considered typical aspects in the ASD phenotype.[398,399] From a mechanists stand point, SCFAs are considered important epigenetic modulators of gene expression, since they are able to promote histones hyperacetylation through the inhibition of histone deacetylases (HDACs), which, in turn, play fundamental roles in brain physiology and disorders, such as Alzheimer's disease, schizophrenia, stress, depression, and addiction.[400] Interestingly, persistent SCFA-mediated HDAC inhibition, through the chronic administration of butyrate, exerted an antidepressant-like effect, inducing a significant decrease of behavioral despair[401] and a durable diminution of fear in mice.[402] To affect the host and brain functions, SCFAs can act also as endogenous ligands for several G protein-coupled receptors (GPCRs), principally, FFAR2 and 3 FFAR3.[403,404] Expression of both receptors has been reported in the colon, in various immune cells, and in the heart. FFAR3, but not FFAR2, is also present in the ANS and CNS, particularly on the BBB.[405,406] In GF mice, the BBB permeability is impaired with respect to conventional control animals and may be ameliorated after systemic treatment with butyrate and propionate.[407] In the colon, through the activation of FFAR2 and FFAR3, SCFAs can stimulate the release of neuropeptides and hormones from EECs, such as the PYY or GLP1, which, besides behaving as anorexing hormones may affect brain circuits involved in behavioral and emotive states.[15,343] In particular, GLP1 takes part in the processes of learning and memory, neuroprotection and neuroplasticity in the hippocampus,[408] and it also seems to be related to reduced β-amyloid plaques and microglia activation in animal models of Alzheimer disease.[409] Analogously, PYY may influence brain functions and affective states, as demonstrated by the presence of aggravated depressive-like behaviors in the $PYYg^{-/-}$ mice.[410]

Finally, in the context of the complex array of cellular players involved in the gut-brain communication, SCFAs may also affect immune cell responses, both by interacting with specific receptors and by activating epigenetic mechanisms. In this view SCFAs have been shown to take part in the regulation of differentiation, recruitment and activation of neutrophils, macrophages, dendritic cells and monocytes.[411] For example, in rodent models, oral administration of butyrate and propionate promoted peripheral regulatory T cell generation.[412]

5.5 Bile acids

Among the different metabolites derived from host–gut microbiota network, bile acids represent one of the best-known class of molecules. In the host organism, bile acids are first synthesized from cholesterol in the liver and then converted into secondary forms by microbial modifications in the intestine.[413] Such microbial modifications enable bile acids to readily modulate signaling and metabolic host processes, such as systemic lipid, cholesterol and glucose metabolism, as well as energy and immune homeostasis via nuclear hormone farnesoid X receptor (FXR) and the G protein-coupled bile acid receptor 1 (TGR5) activation in the gut, liver and periphery.[414] Different studies have shown that $Trg5^{-/-}$ deficient mice had decreased bile acid pool,[415] and the intestinal-specific *Fxr* deletion may protect mice against liver steatosis and obesity, when fed with a high-fat diet.[416,417] In the small intestine, bile acids promote the release of potent regulators of glucose metabolism, such as FGF19 and the incretin GLP-1, thus contributing to the antidiabetic effects of bile acids.[418] Bile acids and gut microbiota are in constant bidirectional communication, influencing each other's composition and potentially leading to neuronal modulation.[6]

Bile acids, indeed, help to limit the expansion of the bacterial population, participating also to the regulation of barrier function and immune status within the gastrointestinal tract. Luminal bile acids deficiency has been related to limited small intestine bacterial overgrowth, activation of inflammation, and gut epithelium damage.[419] This suggest an important role for bile acids as components of intestinal antimicrobial defense, exerting both a direct effect through their membrane-solubilizing properties,[420] and activating different receptor signaling pathways. In the small intestine, for instance, FXR may regulate the expression of antimicrobial defense genes, such as iNOS and IL-18, ensuring the protection from epithelial deterioration and bacterial translocation.[419,421] Interestingly, different studies have reported that the FXR activation is reflected by reduced severity of experimentally- induced colitis in mice.[422,423] In the same way, TGR5 is equally required to support the epithelial barrier and control the inflammatory response in the intestine through, at least in part, the direct action on the GALT.[424,425] *Trg5*$^{-/-}$ mice showed an increased intestinal permeability with respect controls,[424] and the addition of several TGR5 activating compounds led to attenuated production of pro-inflammatory cytokines and inflammatory processes in mouse models of colitis.[426–428] Through their influence on gut microbiota composition and inflammatory pathways, bile acids may have an important impact on the gut-brain communication signaling. In general, maintenance of epithelial integrity is extremely important not only for the gut but also for the brain health. Recently, a correlation between deficient bile acid deconjugation and impaired behavior and intestinal barrier function was found in a mouse model of ASD.[283] This evidence was further related to a substantial reduction in the relative abundance of bile-metabolizing bacteria. However, recent evidences suggest that bile acids can also directly affect the brain function. Interestingly, in the hypothalamus of mice treated with bile acids, by oral gavage, mRNA expression of agouti-related peptide, *AgRP*, was suppressed, leading to glucose tolerance in obese mice.[429] Overall, these findings point to bile acids as microbiota-gut-brain regulatory pathways sustaining host health.

6. Conclusions

The enteric microbiota has a high ability to undergo adaptive changes in response to the host lifestyle (caused by diet, drugs, social, ethnic and environmental factors), strengthening the hypothesis that our behavior may profoundly affect this symbiotic organ. Strong evidence, obtained in the last decades, suggests that the host may communicate with the microbiota by releasing neuroactive molecules, which may impact on microbes. Conversely, the commensal bacteria in our body may interfere with gut and brain functions by releasing bioactive metabolites, via humoral, endocrine, immune and neuronal pathways. In this latter context, it is now widely acknowledged that microbes may produce neuroactive molecules such as neurotransmitters (i.e., noradrenaline, dopamine, serotonin, GABA and glutamate) and metabolites (i.e., tryptophan metabolites, SCFA, secondary bile acids) involved in the regulation of interkingdom communication pathways between eukaryota and prokaryota. In this perspective, accruing experimental evidence from preclinical and clinical studies show that the microbial flora is critical in the development and maintenance of gut and brain function. Thus, alterations of the microflora homeostasis are proposed to participate to both gastrointestinal diseases, often associated with psychiatric disorders, such as IBS and IBD, as well as to CNS disorders such as neurodevelopmental, psychiatric and neurodegenerative diseases. However, considerably more data are needed, especially in the clinical setting, to establish a causative role for such changes and development of the majority of these gut and CNS disorders, since only IBS represents, up to now, a clinical condition where targeting the microbiota may ameliorate symptoms in placebo-controlled trials.[6] An important issue in the physiological balance of the enteric microbiota in the human body is the possibility to define interindividual differences in microbiota composition, allowing personalized therapeutic approaches. In this view, a promising strategy will be to combine different methodologies, including metabolomic, metagenomic, metatranscriptomic and proteomic approaches, in order to identify bacteria and bacterial genes involved in the cross talk with the host and to verify whether their manipulation could be of value as adjuvant of conventional therapies of gut-brain axis related disorders. In particular, research in this field opens an exciting scenario on the possibility target pre- and probiotics producing neuroendocrine neuroactive and immunomodulatory molecules as new potential therapeutic tools addressed to the treatment of neurogastrointestinal and/or psychiatric disorders.

References

1. Baj A, Moro E, Bistoletti M, Orlandi V, Crema F, Giaroni C. Glutamatergic signaling along the microbiota-gut-brain axis. *Int J Mol Sci*. 2019;20(6):1482. https://doi.org/10.3390/ijms20061482.
2. Rhee SH, Pothoulakis C, Mayer EA. Principles and clinical implications of the brain-gut-enteric microbiota axis. *Nat Rev Gastroenterol Hepatol*. 2009;6(5):306–314. https://doi.org/10.1038/nrgastro.2009.35.

3. Julio-Pieper M, Bravo JA. Intestinal barrier and behavior. *Int Rev Neurobiol.* 2016;131:127–141. https://doi.org/10.1016/bs.irn.2016.08.006.
4. Obrenovich MEM. Leaky gut, leaky brain? *Microorganisms.* 2018;6(4):107. https://doi.org/10.3390/microorganisms6040107.
5. Bhattarai Y, Muniz Pedrogo DA, Kashyap PC. Irritable bowel syndrome: a gut microbiota-related disorder? *Am J Physiol Gastrointest Liver Physiol.* 2017;312(1):G52–G62. https://doi.org/10.1152/ajpgi.00338.2016.
6. Cryan JF, O'riordan KJ, Cowan CSM, et al. The microbiota-gut-brain axis. *Physiol Rev.* 2019;99(4):1877–2013. https://doi.org/10.1152/physrev.00018.2018.
7. Sekirov I, Russell SL, Caetano M, Antunes L, Finlay BB. Gut microbiota in health and disease. *Physiol Rev.* 2010;90(3):859–904. https://doi.org/10.1152/physrev.00045.2009.
8. Kelly JR, Minuto C, Cryan JF, Clarke G, Dinan TG. Cross talk: the microbiota and neurodevelopmental disorders. *Front Neurosci.* 2017;11:490. https://doi.org/10.3389/fnins.2017.00490.
9. Bastiaanssen TFS, Cowan CSM, Claesson MJ, Dinan TG, Cryan JF. Making sense of … the microbiome in psychiatry. *Int J Neuropsychopharmacol.* 2019;22(1):37–52. https://doi.org/10.1093/ijnp/pyy067.
10. Pascale A, Marchesi N, Govoni S, Coppola A, Gazzaruso C. The role of gut microbiota in obesity, diabetes mellitus, and effect of metformin: new insights into old diseases. *Curr Opin Pharmacol.* 2019;49:1–5. https://doi.org/10.1016/j.coph.2019.03.011.
11. Cerdó T, Diéguez E, Campoy C. Impact of gut microbiota on neurogenesis and neurological diseases during infancy. *Curr Opin Pharmacol.* 2020;50:33–37. https://doi.org/10.1016/j.coph.2019.11.006.
12. Ni J, Wu GD, Albenberg L, Tomov VT. Gut microbiota and IBD: causation or correlation? *Nat Rev Gastroenterol Hepatol.* 2017;14(10):573–584. https://doi.org/10.1038/nrgastro.2017.88.
13. O'Mahony SM, Clarke G, Borre YE, Dinan TG, Cryan JF. Serotonin, tryptophan metabolism and the brain-gut-microbiome axis. *Behav Brain Res.* 2015;277:32–48. https://doi.org/10.1016/j.bbr.2014.07.027.
14. Mittal R, Debs LH, Patel AP, et al. Neurotransmitters: the critical modulators regulating gut–brain axis. *J Cell Physiol.* 2017;232(9):2359–2372. https://doi.org/10.1002/jcp.25518.
15. Holzer P, Farzi A. Neuropeptides and the microbiota—gut-brain axis. *Adv Exp Med Biol.* 2014;817:196–219. https://doi.org/10.1007/978-1-4939-0897-4_9.
16. Gill SR, Pop M, Deboy RT, et al. Metagenomic analysis of the human distal gut microbiome. *Science (New York, NY).* 2006;312(5778):1355–1359. https://doi.org/10.1126/science.1124234.
17. Qin J, Li R, Raes J, et al. A human gut microbial gene catalogue established by metagenomic sequencing. *Nature.* 2010;464(7285):59–65. https://doi.org/10.1038/nature08821.
18. Ley RE, Turnbaugh PJ, Klein S, Gordon JI. Microbial ecology: human gut microbes associated with obesity. *Nature.* 2006;444(7122):1022–1023. https://doi.org/10.1038/4441022a.
19. Sender R, Fuchs S, Milo R. Revised estimates for the number of human and bacteria cells in the body. *PLoS Biol.* 2016;14(8):e1002533. https://doi.org/10.1371/journal.pbio.1002533.
20. Deng P, Swanson KS. Gut microbiota of humans, dogs and cats: current knowledge and future opportunities and challenges. *Br J Nutr.* 2015;113(Suppl):S6–S17. https://doi.org/10.1017/S0007114514002943.
21. Wang Q, Wang K, Wu W, Giannoulatou E, Ho JWK, Li L. Host and microbiome multi-omics integration: applications and methodologies. *Biophys Rev.* 2019;11:55–65. https://doi.org/10.1007/s12551-018-0491-7.

22. Lauder AP, Roche AM, Sherrill-Mix S, et al. Comparison of placenta samples with contamination controls does not provide evidence for a distinct placenta microbiota. *Microbiome*. 2016;4:29. https://doi.org/10.1186/s40168-016-0172-3.
23. Aagaard K, Ma J, Antony KM, Ganu R, Petrosino J, Versalovic J. The placenta harbors a unique microbiome. *Sci Transl Med*. 2014;6(237):237ra65. https://doi.org/10.1126/scitranslmed.3008599.
24. DiGiulio DB. Diversity of microbes in amniotic fluid. *Semin Fetal Neonatal Med*. 2012;17(1):2–11. https://doi.org/10.1016/j.siny.2011.10.001.
25. Kuperman AA, Koren O. Antibiotic use during pregnancy: how bad is it? *BMC Med*. 2016;14(1):91. https://doi.org/10.1186/s12916-016-0636-0.
26. Sharon G, Sampson TR, Geschwind DH, Mazmanian SK. The central nervous system and the gut microbiome. *Cell*. 2016;167(4):915–932. https://doi.org/10.1016/j.cell.2016.10.027.
27. Dominguez-Bello MG, Costello EK, Contreras M, et al. Delivery mode shapes the acquisition and structure of the initial microbiota across multiple body habitats in newborns. *Proc Natl Acad Sci U S A*. 2010;107(26):11971–11975. https://doi.org/10.1073/pnas.1002601107.
28. Bäckhed F, Roswall J, Peng Y, et al. Dynamics and stabilization of the human gut microbiome during the first year of life. *Cell Host Microbe*. 2015;17(5):690–703. https://doi.org/10.1016/j.chom.2015.04.004.
29. Penders J, Thijs C, Vink C, et al. Factors influencing the composition of the intestinal microbiota in early infancy. *Pediatrics*. 2006;118(2):511–521. https://doi.org/10.1542/peds.2005-2824.
30. Neu J, Walker WA. Necrotizing enterocolitis. *N Engl J Med*. 2011;364(3):255–264. https://doi.org/10.1056/NEJMra1005408.
31. Fernández L, Langa S, Martín V, et al. The human milk microbiota: origin and potential roles in health and disease. *Pharmacol Res*. 2013;69(1):1–10. https://doi.org/10.1016/j.phrs.2012.09.001.
32. Hill DR, Rho HK, Kessler SP, et al. Human milk hyaluronan enhances innate defense of the intestinal epithelium. *J Biol Chem*. 2013;288(40):29090–29104. https://doi.org/10.1074/jbc.M113.468629.
33. Palmer C, Bik EM, DiGiulio DB, Relman DA, Brown PO. Development of the human infant intestinal microbiota. *PLoS Biol*. 2007;5(7):1556–1573. https://doi.org/10.1371/journal.pbio.0050177.
34. Lloyd-Price J, Abu-Ali G, Huttenhower C. The healthy human microbiome. *Genome Med*. 2016;8(1):51. https://doi.org/10.1186/s13073-016-0307-y.
35. Borre YE, O'Keeffe GW, Clarke G, Stanton C, Dinan TG, Cryan JF. Microbiota and neurodevelopmental windows: implications for brain disorders. *Trends Mol Med*. 2014;20:509–518. https://doi.org/10.1016/j.molmed.2014.05.002.
36. Rodríguez JM, Murphy K, Stanton C, et al. The composition of the gut microbiota throughout life, with an emphasis on early life. *Microb Ecol Health Dis*. 2015;26:26050. https://doi.org/10.3402/mehd.v26.26050.
37. Fujio-Vejar S, Vasquez Y, Morales P, et al. The gut microbiota of healthy Chilean subjects reveals a high abundance of the phylum Verrucomicrobia. *Front Microbiol*. 2017;8(JUN):1221. https://doi.org/10.3389/fmicb.2017.01221.
38. López-Otín C, Blasco MA, Partridge L, Serrano M, Kroemer G. The hallmarks of aging. *Cell*. 2013;153(6):1194. https://doi.org/10.1016/j.cell.2013.05.039.
39. Biagi E, Franceschi C, Rampelli S, Capri M, Brigidi P, Candela M. Gut microbiota and extreme longevity. *Curr Biol*. 2016;26:1480–1485. https://doi.org/10.1016/j.cub.2016.04.016.
40. O'Toole PW, Claesson MJ. Gut microbiota: changes throughout the lifespan from infancy to elderly. *Int Dairy J*. 2010;20(4):281–291. https://doi.org/10.1016/j.idairyj.2009.11.010.

41. Salazar N, Arboleya S, Valdés L, et al. The human intestinal microbiome at extreme ages of life. Dietary intervention as a way to counteract alterations. *Front Genet.* 2014;5(NOV):406. https://doi.org/10.3389/fgene.2014.00406.
42. Claesson MJ, Cusack S, O'Sullivan O, et al. Composition, variability, and temporal stability of the intestinal microbiota of the elderly. *Proc Natl Acad Sci U S A.* 2011; 108(suppl 1):4586–4591. https://doi.org/10.1073/pnas.1000097107.
43. Mariat D, Firmesse O, Levenez F, et al. The firmicutes/bacteroidetes ratio of the human microbiota changes with age. *BMC Microbiol.* 2009;9:123. https://doi.org/10.1186/1471-2180-9-123.
44. Xia S, Zhang X, Zheng S, et al. An update on inflamm-aging: mechanisms, prevention, and treatment. *J Immunol Res.* 2016;2016:8426874. https://doi.org/10.1155/2016/8426874.
45. Thevaranjan N, Puchta A, Schulz C, et al. Age-associated microbial dysbiosis promotes intestinal permeability, systemic inflammation, and macrophage dysfunction. *Cell Host Microbe.* 2017;21(4):455–466.e4. https://doi.org/10.1016/j.chom.2017.03.002.
46. Hopkins MJ, Sharp R, Macfarlane GT. Age and disease related changes in intestinal bacterial populations assessed by cell culture, 16S rRNA abundance, and community cellular fatty acid profiles. *Gut.* 2001;48(2):198–205. https://doi.org/10.1136/gut.48.2.198.
47. Van Tongeren SP, Slaets JPJ, Harmsen HJM, Welling GW. Fecal microbiota composition and frailty. *Appl Environ Microbiol.* 2005;71(10):6438–6442. https://doi.org/10.1128/AEM.71.10.6438-6442.2005.
48. Zwielehner J, Liszt K, Handschur M, Lassl C, Lapin A, Haslberger AG. Combined PCR-DGGE fingerprinting and quantitative-PCR indicates shifts in fecal population sizes and diversity of bacteroides, bifidobacteria and clostridium cluster IV in institutionalized elderly. *Exp Gerontol.* 2009;44(6–7):440–446. https://doi.org/10.1016/j.exger.2009.04.002.
49. Scott KA, Ida M, Peterson VL, et al. Revisiting Metchnikoff: age-related alterations in microbiota-gut-brain axis in the mouse. *Brain Behav Immun.* 2017;65:20–32. https://doi.org/10.1016/j.bbi.2017.02.004.
50. Eloe-Fadrosh EA, Brady A, Crabtree J, et al. Functional dynamics of the gut microbiome in elderly people during probiotic consumption. *MBio.* 2015;6(2):e00231-15. https://doi.org/10.1128/mBio.00231-15.
51. Smith P, Willemsen D, Popkes M, et al. Regulation of life span by the gut microbiota in the short-lived african turquoise killifish. *Elife.* 2017;6:e27014. https://doi.org/10.7554/eLife.27014.
52. Rowland I, Gibson G, Heinken A, et al. Gut microbiota functions: metabolism of nutrients and other food components. *Eur J Nutr.* 2018;57(1):1–24. https://doi.org/10.1007/s00394-017-1445-8.
53. Bergman EN. Energy contributions of volatile fatty acids from the gastrointestinal tract in various species. *Physiol Rev.* 1990;70(2):567–590. https://doi.org/10.1152/physrev.1990.70.2.567.
54. Sumi Y, Miyakawa M, Kanzaki M, Kotake Y. Vitamin B-6 deficiency in germfree rats. *J Nutr.* 1977;107(9):1707–1714. https://doi.org/10.1093/jn/107.9.1707.
55. Wostmann BS, Larkin C, Moriarty A, Bruckner-Kardoss E. Dietary intake, energy metabolism, and excretory losses of adult male germfree Wistar rats. *Lab Anim Sci.* 1983;33(1):46–50.
56. Macfarlane S, Macfarlane GT. Regulation of short-chain fatty acid production. *Proc Nutr Soc.* 2003;62(1):67–72. https://doi.org/10.1079/pns2002207.
57. Rooks MG, Garrett WS. Gut microbiota, metabolites and host immunity. *Nat Rev Immunol.* 2016;16(6):341–352. https://doi.org/10.1038/nri.2016.42.

58. Kitamoto S, Nagao-Kitamoto H, Kuffa P, Kamada N. Regulation of virulence: the rise and fall of gastrointestinal pathogens. *J Gastroenterol*. 2016;51(3):195–205. https://doi.org/10.1007/s00535-015-1141-5.
59. Martin CR, Osadchiy V, Kalani A, Mayer EA. The brain-gut-microbiome axis. *Cell Mol Gastroenterol Hepatol*. 2018;6(2):133–148. https://doi.org/10.1016/j.jcmgh.2018.04.003.
60. Li Q, Ren Y, Fu X. Inter-kingdom signaling between gut microbiota and their host. *Cell Mol Life Sci*. 2019;76:2382–2389. https://doi.org/10.1007/s00018-019-03076-7.
61. Kaper JB, Sperandio V. Bacterial cell-to-cell signaling in the gastrointestinal tract. *Infect Immun*. 2005;73(6):3197–3209. https://doi.org/10.1128/IAI.73.6.3197-3209.2005.
62. Sintim HO, Smith JA, Wang J, Nakayama S, Yan L. Paradigm shift in discovering next-generation anti-infective agents: targeting quorum sensing, c-di-GMP signaling and biofilm formation in bacteria with small molecules. *Future Med Chem*. 2010; 2(6):1005–1035. https://doi.org/10.4155/fmc.10.185.
63. Thompson JA, Oliveira RA, Djukovic A, Ubeda C, Xavier KB. Manipulation of the quorum sensing signal AI-2 affects the antibiotic-treated gut microbiota. *Cell Rep*. 2015;10(11):1861–1871. https://doi.org/10.1016/j.celrep.2015.02.049.
64. Galloway WRJD, Hodgkinson JT, Bowden SD, Welch M, Spring DR. Quorum sensing in gram-negative bacteria: small-molecule modulation of AHL and AI-2 quorum sensing pathways. *Chem Rev*. 2011;111(1):28–67. https://doi.org/10.1021/cr100109t.
65. Ismail AS, Valastyan JS, Bassler BL. A host-produced autoinducer-2 mimic activates bacterial quorum sensing. *Cell Host Microbe*. 2016;19(4):470–480. https://doi.org/10.1016/j.chom.2016.02.020.
66. Hughes DT, Sperandio V. Inter-kingdom signalling: communication between bacteria and their hosts. *Nat Rev Microbiol*. 2008;6(2):111–120. https://doi.org/10.1038/nrmicro1836.
67. Elmanfi S, Ma X, Sintim HO, Könönen E, Syrjänen S, Gursoy UK. Quorum-sensing molecule dihydroxy-2,3-pentanedione and its analogs as regulators of epithelial integrity. *J Periodontal Res*. 2018;53(3):414–421. https://doi.org/10.1111/jre.12528.
68. Zargar A, Quan DN, Carter KK, et al. Bacterial secretions of nonpathogenic *Escherichia coli* elicit inflammatory pathways: a closer investigation of interkingdom signaling. *MBio*. 2015;6(2):e00025. https://doi.org/10.1128/mBio.00025-15.
69. Xavier KB, Bassler BL. Interference with AI-2-mediated bacterial cell-cell communication. *Nature*. 2005;437(7059):750–753. https://doi.org/10.1038/nature03960.
70. Lesouhaitier O, Clamens T, Rosay T, et al. Host peptidic hormones affecting bacterial biofilm formation and virulence. *J Innate Immun*. 2019;11(3):227–241. https://doi.org/10.1159/000493926.
71. Vidal JE, Howery KE, Ludewick HP, Nava P, Klugman KP. Quorum-sensing systems LuxS/autoinducer 2 and com regulate *Streptococcus pneumoniae* biofilms in a bioreactor with living cultures of human respiratory cells. *Infect Immun*. 2013;81(4):1341–1353. https://doi.org/10.1128/IAI.01096-12.
72. Dejea CM, Wick EC, Hechenbleikner EM, et al. Microbiota organization is a distinct feature of proximal colorectal cancers. *Proc Natl Acad Sci U S A*. 2014;111 (51):18321–18326. https://doi.org/10.1073/pnas.1406199111.
73. Scheres N, Lamont RJ, Crielaard W, Krom BP. LuxS signaling in porphyromonas gingivalis-host interactions. *Anaerobe*. 2015;35:3–9. https://doi.org/10.1016/j.anaerobe.2014.11.011.
74. Yu J, Chen Y, Fu X, et al. Invasive fusobacterium nucleatum may play a role in the carcinogenesis of proximal colon cancer through the serrated neoplasia pathway. *Int J Cancer*. 2016;139(6):1318–1326. https://doi.org/10.1002/ijc.30168.

75. Kau AL, Ahern PP, Griffin NW, Goodman AL, Gordon JI. Human nutrition, the gut microbiome and the immune system. *Nature.* 2011;474(7351):327–336. https://doi.org/10.1038/nature10213.
76. Tateda K, Ishii Y, Horikawa M, et al. The *Pseudomonas aeruginosa* autoinducer N-3-oxododecanoyl homoserine lactone accelerates apoptosis in macrophages and neutrophils. *Infect Immun.* 2003;71(10):5785–5793. https://doi.org/10.1128/IAI.71.10.5785-5793.2003.
77. Shiner EK, Terentyev D, Bryan A, et al. *Pseudomonas aeruginosa* autoinducer modulates host cell responses through calcium signalling. *Cell Microbiol.* 2006;8(10):1601–1610. https://doi.org/10.1111/j.1462-5822.2006.00734.x.
78. Telford G, Wheeler D, Williams P, et al. The *Pseudomonas aeruginosa* quorum-sensing signal molecule N-(3-oxododecanoyl)-L-homoserine lactone has immunomodulatory activity. *Infect Immun.* 1998;66(1):36–42. https://doi.org/10.1128/iai.66.1.36-42.1998.
79. Ritchie AJ, Yam AOW, Tanabe KM, Rice SA, Cooley MA. Modification of in vivo and in vitro T- and B-cell-mediated immune responses by the *Pseudomonas aeruginosa* quorum-sensing molecule n-(3-oxododecanoyl)-L-homoserine lactone. *Infect Immun.* 2003;71(8):4421–4431. https://doi.org/10.1128/IAI.71.8.4421-4431.2003.
80. Chhabra SR, Harty C, Hooi DSW, et al. Synthetic analogues of the bacterial signal (quorum sensing) molecule N-(3-oxododecanoyl)-L-homoserine lactone as immune modulators. *J Med Chem.* 2003;46(1):97–104. https://doi.org/10.1021/jm020909n.
81. Yeo S, Park H, Ji Y, et al. Influence of gastrointestinal stress on autoinducer-2 activity of two *Lactobacillus* species. Nakatsu C, Ed, *FEMS Microbiol Ecol.* 2015;91(7):fiv065. https://doi.org/10.1093/femsec/fiv065.
82. Williams SC, Patterson EK, Carty NL, Griswold JA, Hamood AN, Rumbaugh KP. Pseudomonas aeruginosa autoinducer enters and functions in mammalian cells. *J Bacteriol.* 2004;186(8):2281–2287. https://doi.org/10.1128/JB.186.8.2281-2287.2004.
83. Sperandio V, Torres AG, Jarvis B, Nataro JP, Kaper JB. Bacteria-host communication: the language of hormones. *Proc Natl Acad Sci U S A.* 2003;100(15):8951–8956. https://doi.org/10.1073/pnas.1537100100.
84. Lyte M, Frank CD, Green BT. Production of an autoinducer of growth by norepinephrine cultured *Escherichia coli* O157:H7. *FEMS Microbiol Lett.* 1996;139(2–3):155–159. https://doi.org/10.1111/j.1574-6968.1996.tb08196.x.
85. Moreira CG, Sperandio V. The epinephrine/norepinephrine/autoinducer-3 interkingdom signaling system in Escherichia coli o157:H7. *Adv Exp Med Biol.* 2016;874:247–261. https://doi.org/10.1007/978-3-319-20215-0_12.
86. Deng Y, Li M, Mei L, et al. Manipulation of intestinal dysbiosis by a bacterial mixture ameliorates loperamide-induced constipation in rats. *Benefic Microbes.* 2018;9(3):453–464. https://doi.org/10.3920/BM2017.0062.
87. Giaroni C, De Ponti F, Cosentino M, Lecchini S, Frigo G. Plasticity in the enteric nervous system. *Gastroenterology.* 1999;117(6):1438–1458.
88. Ramanan D, Cadwell K. Intrinsic defense mechanisms of the intestinal epithelium. *Cell Host Microbe.* 2016;19(4):434–441. https://doi.org/10.1016/j.chom.2016.03.003.
89. Chelakkot C, Ghim J, Ryu SH. Mechanisms regulating intestinal barrier integrity and its pathological implications. *Exp Mol Med.* 2018;50(8):1–9. https://doi.org/10.1038/s12276-018-0126-x.
90. Jandhyala SM, Talukdar R, Subramanyam C, Vuyyuru H, Sasikala M, Reddy DN. Role of the normal gut microbiota. *World J Gastroenterol.* 2015;21(29):8836–8847. https://doi.org/10.3748/wjg.v21.i29.8787.
91. Arike L, Seiman A, van der Post S, et al. Protein turnover in epithelial cells and mucus along the gastrointestinal tract is coordinated by the spatial location and microbiota. *Cell Rep.* 2020;30(4):1077–1087.e3. https://doi.org/10.1016/j.celrep.2019.12.068.

92. Hooper LV, Wong MH, Thelin A, Hansson L, Falk PG, Gordon JI. Molecular analysis of commensal host-microbial relationships in the intestine. *Science*. 2001;291 (5505):881–884. https://doi.org/10.1126/science.291.5505.881.
93. Yan F, Cao H, Cover TL, et al. Colon-specific delivery of a probiotic-derived soluble protein ameliorates intestinal inflammation in mice through an EGFR-dependent mechanism. *J Clin Investig*. 2011;121(6):2242–2253. https://doi.org/10.1172/JCI44031.
94. Yan F, Polk DB. Probiotic bacterium prevents cytokine-induced apoptosis in intestinal epithelial cells. *J Biol Chem*. 2002;277(52):50959–50965. https://doi.org/10.1074/jbc.M207050200.
95. Hughes KR, Harnisch LC, Alcon-Giner C, et al. Bifidobacterium breve reduces apoptotic epithelial cell shedding in an exopolysaccharide and MyD88-dependent manner. *Open Biol*. 2017;7(1):160155. https://doi.org/10.1098/rsob.160155.
96. Abrams GD, Bauer H, Sprinz H. Influence of the normal flora on mucosal morphology and cellular renewal in the ileum. A comparison of germ-free and conventional mice. *Lab Invest*. 1963;12:355–364.
97. Caputi V, Giron MC. Microbiome-gut-brain axis and toll-like receptors in parkinson's disease. *Int J Mol Sci*. 2018;19(6):1689. https://doi.org/10.3390/ijms19061689.
98. Kawasaki T, Kawai T. Toll-like receptor signaling pathways. *Front Immunol*. 2014;5 (SEP):461. https://doi.org/10.3389/fimmu.2014.00461.
99. Takeuchi O, Akira S. Pattern recognition receptors and inflammation. *Cell*. 2010;140 (6):805–820. https://doi.org/10.1016/j.cell.2010.01.022.
100. Rakoff-Nahoum S, Paglino J, Eslami-Varzaneh F, Edberg S, Medzhitov R. *Recognition of Commensal Microflora by Toll-Like Receptors is Required for Intestinal Homeostasis*. vol 118; 2004. https://doi.org/10.1016/j.cell.2004.07.002.
101. Cario E, Gerken G, Podolsky DK. Toll-like receptor 2 enhances ZO-1-associated intestinal epithelial barrier integrity via protein kinase C. *Gastroenterology*. 2004;127 (1):224–238. https://doi.org/10.1053/j.gastro.2004.04.015.
102. Ayabe T, Satchell DP, Wilson CL, Parks WC, Selsted ME, Ouellette AJ. Secretion of microbicidal α-defensins by intestinal paneth cells in response to bacteria. *Nat Immunol*. 2000;1(2):113–118. https://doi.org/10.1038/77783.
103. Vaishnava S, Behrendt CL, Ismail AS, Eckmann L, Hooper LV. Paneth cells directly sense gut commensals and maintain homeostasis at the intestinal host-microbial interface. *Proc Natl Acad Sci U S A*. 2008;105(52):20858–20863. https://doi.org/10.1073/pnas.0808723105.
104. Hooper LV, Stappenbeck TS, Hong CV, Gordon JI. Angiogenins: a new class of microbicidal proteins involved in innate immunity. *Nat Immunol*. 2003;4(3):269–273. https://doi.org/10.1038/ni888.
105. Cash HL, Whitham CV, Behrendt CL, Hooper LV. Symbiotic bacteria direct expression of an intestinal bactericidal lectin. *Science (New York, NY)*. 2006;313 (5790):1126–1130. https://doi.org/10.1126/science.1127119.
106. Corr SC, Gahan CGM, Hill C. Impact of selected *Lactobacillus* and *Bifidobacterium* species on *Listeria monocytogenes* infection and the mucosal immune response. *FEMS Immunol Med Microbiol*. 2007;50(3):380–388. https://doi.org/10.1111/j.1574-695X.2007.00264.x.
107. Corr SC, Li Y, Riedel CU, O'Toole PW, Hill C, Gahan CGM. Bacteriocin production as a mechanism for the antiinfective activity of lactobacillus salivarius UCC118. *Proc Natl Acad Sci U S A*. 2007;104(18):7617–7621. https://doi.org/10.1073/pnas.0700440104.
108. Gareau MG, Silva MA, Perdue MH. Pathophysiological mechanisms of stress-induced intestinal damage. *Curr Mol Med*. 2008;8(4):274–281. https://doi.org/10.2174/156652408784533760.

109. Maes M, Kubera M, Leunis JC, Berk M. Increased IgA and IgM responses against gut commensals in chronic depression: further evidence for increased bacterial translocation or leaky gut. *J Affect Disord.* 2012;141(1):55–62. https://doi.org/10.1016/j.jad.2012.02.023.
110. Maes M, Kubera M, Leunis JC. The gut-brain barrier in major depression: intestinal mucosal dysfunction with an increased translocation of LPS from gram negative enterobacteria (leaky gut) plays a role in the inflammatory pathophysiology of depression. *Neuro Endocrinol Lett.* 2008;29(1):117–124.
111. Maes M, Leunis JC. Normalization of leaky gut in chronic fatigue syndrome (CFS) is accompanied by a clinical improvement: effects of age, duration of illness and the translocation of LPS from gram-negative bacteria. *Neuroendocrinol Lett.* 2008; 29(6):902–910.
112. Sarkar A, Lehto SM, Harty S, Dinan TG, Cryan JF, Burnet PWJ. Psychobiotics and the manipulation of bacteria–gut–brain signals. *Trends Neurosci.* 2016;39(11):763–781. https://doi.org/10.1016/j.tins.2016.09.002.
113. Hsiao EY, McBride SW, Hsien S, et al. Microbiota modulate behavioral and physiological abnormalities associated with neurodevelopmental disorders (La microbiota modula la fisiología intestinal y las anomalías conductuales asociadas con el autismo). *Cell.* 2013;155(7):1451–1463. https://doi.org/10.1016/j.cell.2013.11.024.
114. Ait-Belgnaoui A, Durand H, Cartier C, et al. Prevention of gut leakiness by a probiotic treatment leads to attenuated HPA response to an acute psychological stress in rats. *Psychoneuroendocrinology.* 2012;37(11):1885–1895. https://doi.org/10.1016/j.psyneuen.2012.03.024.
115. Miyauchi E, O'Callaghan J, Buttó LF, et al. Mechanism of protection of transepithelial barrier function by Lactobacillus salivarius: strain dependence and attenuation by bacteriocin production. *Am J Physiol Gastrointest Liver Physiol.* 2012;303(9):G1029–G1041. https://doi.org/10.1152/ajpgi.00003.2012.
116. Sternini C, Anselmi L, Rozengurt E. Enteroendocrine cells: a site of "taste" in gastrointestinal chemosensing. *Curr Opin Endocrinol Diabetes Obes.* 2008;15(1):73–78. https://doi.org/10.1097/MED.0b013e3282f43a73.
117. Fothergill LJ, Furness JB. Diversity of enteroendocrine cells investigated at cellular and subcellular levels: the need for a new classification scheme. *Histochem Cell Biol.* 2018;150(6):693–702. https://doi.org/10.1007/s00418-018-1746-x.
118. Gunawardene AR, Corfe BM, Staton CA. Classification and functions of enteroendocrine cells of the lower gastrointestinal tract. *Int J Exp Pathol.* 2011;92(4):219–231. https://doi.org/10.1111/j.1365-2613.2011.00767.x.
119. Gribble FM, Reimann F. Enteroendocrine cells: chemosensors in the intestinal epithelium. *Annu Rev Physiol.* 2016;78(1):277–299. https://doi.org/10.1146/annurev-physiol-021115-105439.
120. Bellono NW, Bayrer JR, Leitch DB, et al. Enterochromaffin cells are gut chemosensors that couple to sensory neural pathways. *Cell.* 2017;170(1):185–198.e16. https://doi.org/10.1016/j.cell.2017.05.034.
121. Tsubouchi S, Leblond CP. Migration and turnover of entero-endocrine and caveolated cells in the epithelium of the descending colon, as shown by radioautography after continuous infusion of 3H-thymidine into mice. *Am J Anat.* 1979;156(4):431–451. https://doi.org/10.1002/aja.1001560403.
122. Yu Y, Yang W, Li Y, Cong Y. Enteroendocrine cells: sensing gut microbiota and regulating inflammatory bowel diseases. *Inflamm Bowel Dis.* 2020;26(1):11–20. https://doi.org/10.1093/ibd/izz217.
123. Habib AM, Richards P, Rogers GJ, Reimann F, Gribble FM. Co-localisation and secretion of glucagon-like peptide 1 and peptide YY from primary cultured human L cells. *Diabetologia.* 2013;56(6):1413–1416. https://doi.org/10.1007/s00125-013-2887-z.

124. Habib AM, Richards P, Cairns LS, et al. Overlap of endocrine hormone expression in the mouse intestine revealed by transcriptional profiling and flow cytometry. *Endocrinology*. 2012;153(7):3054–3065. https://doi.org/10.1210/en.2011-2170.
125. Steinert RE, Feinle-Bisset C, Asarian L, Horowitz M, Beglinger C, Geary N. Ghrelin, CCK, GLP-1, and PYY(3–36): secretory controls and physiological roles in eating and glycemia in health, obesity, and after RYGB. *Physiol Rev*. 2017;97(1):411–463. https://doi.org/10.1152/physrev.00031.2014.
126. De Silva A, Bloom SR. Gut hormones and appetite control: a focus on PYY and GLP-1 as therapeutic targets in obesity. *Gut Liver*. 2012;6(1):10–20. https://doi.org/10.5009/gnl.2012.6.1.10.
127. Richards P, Parker HE, Adriaenssens AE, et al. Identification and characterization of GLP-1 receptor-expressing cells using a new transgenic mouse model. *Diabetes*. 2014;63(4):1224–1233. https://doi.org/10.2337/db13-1440.
128. Kaelberer MM, Buchanan KL, Klein ME, et al. A gut-brain neural circuit for nutrient sensory transduction. *Science (New York, NY)*. 2018;361(6408):eaat5236. https://doi.org/10.1126/science.aat5236.
129. Engelstoft MS, Egerod KL, Holst B, Schwartz TW. A gut feeling for obesity: 7TM sensors on enteroendocrine cells. *Cell Metab*. 2008;8(6):447–449. https://doi.org/10.1016/j.cmet.2008.11.004.
130. Lu VB, Gribble FM, Reimann F. Free fatty acid receptors in enteroendocrine cells. *Endocrinology*. 2018;159(7):2826–2835. https://doi.org/10.1210/en.2018-00261.
131. Gorboulev V, Schürmann A, Vallon V, et al. Na(+)-D-glucose cotransporter SGLT1 is pivotal for intestinal glucose absorption and glucose-dependent incretin secretion. *Diabetes*. 2012;61(1):187–196. https://doi.org/10.2337/db11-1029.
132. Spreckley E, Murphy KG. The L-cell in nutritional sensing and the regulation of appetite. *Front Nutr*. 2015;2:23. https://doi.org/10.3389/fnut.2015.00023.
133. Nguyen AT, Mandard S, Dray C, et al. Lipopolysaccharides-mediated increase in glucose-stimulated insulin secretion: involvement of the GLP-1 pathway. *Diabetes*. 2014;63(2):471–482. https://doi.org/10.2337/db13-0903.
134. Cani PD, Everard A, Duparc T. Gut microbiota, enteroendocrine functions and metabolism. *Curr Opin Pharmacol*. 2013;13(6):935–940. https://doi.org/10.1016/j.coph.2013.09.008.
135. Larraufie P, Doré J, Lapaque N, Blottière HM. TLR ligands and butyrate increase Pyy expression through two distinct but inter-regulated pathways. *Cell Microbiol*. 2017;19(2):e12648. https://doi.org/10.1111/cmi.12648.
136. Brooks L, Viardot A, Tsakmaki A, et al. Fermentable carbohydrate stimulates FFAR2-dependent colonic PYY cell expansion to increase satiety. *Mol Metab*. 2017;6(1):48–60. https://doi.org/10.1016/j.molmet.2016.10.011.
137. Tolhurst G, Heffron H, Lam YS, et al. Short-chain fatty acids stimulate glucagon-like peptide-1 secretion via the G-protein-coupled receptor FFAR2. *Diabetes*. 2012;61(2):364–371. https://doi.org/10.2337/db11-1019.
138. Katsuma S, Hirasawa A, Tsujimoto G. Bile acids promote glucagon-like peptide-1 secretion through TGR5 in a murine enteroendocrine cell line STC-1. *Biochem Biophys Res Commun*. 2005;329(1):386–390. https://doi.org/10.1016/j.bbrc.2005.01.139.
139. Thomas C, Gioiello A, Noriega L, et al. TGR5-mediated bile acid sensing controls glucose homeostasis. *Cell Metab*. 2009;10(3):167–177. https://doi.org/10.1016/j.cmet.2009.08.001.
140. Gribble FM, Reimann F. Enteroendocrine cells: chemosensors in the intestinal epithelium. *Annu Rev Physiol*. 2016;78(1):277–299. https://doi.org/10.1146/annurev-physiol-021115-105439.

141. Cani PD, Daubioul CA, Reusens B, Remacle C, Catillon G, Delzenne NM. Involvement of endogenous glucagon-like peptide-1(7-36) amide on glycaemia-lowering effect of oligofructose in streptozotocin-treated rats. *J Endocrinol.* 2005;185 (3):457–465. https://doi.org/10.1677/joe.1.06100.
142. Everard A, Lazarevic V, Derrien M, et al. Responses of gut microbiota and glucose and lipid metabolism to prebiotics in genetic obese and diet-induced leptin-resistant mice. *Diabetes.* 2011;60(11):2775–2786. https://doi.org/10.2337/db11-0227.
143. Panwar H, Calderwood D, Gillespie AL, et al. Identification of lactic acid bacteria strains modulating incretin hormone secretion and gene expression in enteroendocrine cells. *J Funct Foods.* 2016;23:348–358. https://doi.org/10.1016/j.jff.2016.02.040.
144. Belguesmia Y, Alard J, Mendil R, et al. In vitro probiotic properties of selected lactobacilli and multi-strain consortium on immune function, gut barrier strengthening and gut hormone secretion. *J Funct Foods.* 2019;57:382–391. https://doi.org/10.1016/j.jff.2019.04.028.
145. Simon MC, Strassburger K, Nowotny B, et al. Intake of lactobacillus reuteri improves incretin and insulin secretion in glucose-tolerant humans: a proof of concept. *Diabetes Care.* 2015;38(10):1827–1834. https://doi.org/10.2337/dc14-2690.
146. Popoff MR, Poulain B. Bacterial toxins and the nervous system: neurotoxins and multipotential toxins interacting with neuronal cells. *Toxins.* 2010;2(4):683–737. https://doi.org/10.3390/toxins2040683.
147. Peregrin AT, Ahlman H, Jodal M, Lundgren O. Involvement of serotonin and calcium channels in the intestinal fluid secretion evoked by bile salt and cholera toxin. *Br J Pharmacol.* 1999;127(4):887–894. https://doi.org/10.1038/sj.bjp.0702615.
148. Stephens RL, Tache Y. Intracisternal injection of a TRH analogue stimulates gastric luminal serotonin release in rats. *Am J Physiol.* 1989;256(2 Pt 1):G377–G383. https://doi.org/10.1152/ajpgi.1989.256.2.G377.
149. Mawe GM, Hoffman JM. Serotonin {signaling} in the {gastrointestinal} {tract}. *Nat Rev Gastroenterol Hepatol.* 2013;10(8):473–486. https://doi.org/10.1038/nrgastro.2013.105.
150. Gershon MD, Tack J. The serotonin signaling system: from basic understanding to drug development for functional GI disorders. *Gastroenterology.* 2007;132(1):397–414. https://doi.org/10.1053/j.gastro.2006.11.002.
151. Hansen MB, Witte A-B. The role of serotonin in intestinal luminal sensing and secretion. *Acta Physiol (Oxf).* 2008;193(4):311–323. https://doi.org/10.1111/j.1748-1716.2008.01870.x.
152. Bogunovic M, Davé SH, Tilstra JS, et al. Enteroendocrine cells express functional toll-like receptors. *Am J Physiol Gastrointest Liver Physiol.* 2007;292(6):G1770–G1783. https://doi.org/10.1152/ajpgi.00249.2006.
153. Lencer WI. Microbes and microbial toxins: paradigms for microbial-mucosal interactions V. Cholera: invasion of the intestinal epithelial barrier by a stably folded protein toxin. *Am J Physiol Gastrointest Liver Physiol.* 2001;280(5):43–45. https://doi.org/10.1152/ajpgi.2001.280.5.G781.
154. Grøndahl ML, Unmack MA, Ragnarsdóttir HB, Hansen MB, Olsen JE, Skadhauge E. Effects of nitric oxide in 5-hydroxytryptamine-, cholera toxin-, enterotoxigenic *Escherichia coli-* and Salmonella typhimurium-induced secretion in the porcine small intestine. *Comp Biochem Physiol A Mol Integr Physiol.* 2005;141(4):476–484. https://doi.org/10.1016/j.cbpb.2005.07.001.
155. Xue J, Askwith C, Javed NH, Cooke HJ. Autonomic nervous system and secretion across the intestinal mucosal surface. *Auton Neurosci.* 2007;133(1):55–63. https://doi.org/10.1016/j.autneu.2007.02.001.

156. Mawe GM, Coates MD, Moses PL. Review article: intestinal serotonin signalling in irritable bowel syndrome. *Aliment Pharmacol Ther.* 2006;23(8):1067–1076. https://doi.org/10.1111/j.1365-2036.2006.02858.x.
157. Wheatcroft J, Wakelin D, Smith A, Mahoney CR, Mawe G, Spiller R. Enterochromaffin cell hyperplasia and decreased serotonin transporter in a mouse model of postinfectious bowel dysfunction. *Neurogastroenterol Motil.* 2005;17(6):863–870. https://doi.org/10.1111/j.1365-2982.2005.00719.x.
158. Donovan MH, Tecott LH. Serotonin and the regulation of mammalian energy balance. *Front Neurosci.* 2013;7:36. https://doi.org/10.3389/fnins.2013.00036.
159. Mawe GM, Hoffman JM. Serotonin signalling in the gut-functions, dysfunctions and therapeutic targets. *Nat Rev Gastroenterol Hepatol.* 2013;10(8):473–486. https://doi.org/10.1038/nrgastro.2013.105.
160. Browning KN. Role of central vagal 5-HT3 receptors in gastrointestinal physiology and pathophysiology. *Front Neurosci.* 2015;9(OCT):413. https://doi.org/10.3389/fnins.2015.00413.
161. Shajib MS, Baranov A, Khan WI. Diverse effects of gut-derived serotonin in intestinal inflammation. *ACS Chem Nerosci.* 2017;8(5):920–931. https://doi.org/10.1021/acschemneuro.6b00414.
162. Round JL, Mazmanian SK. The gut microbiota shapes intestinal immune responses during health and disease. *Nat Rev Immunol.* 2009;9(5):313–323. https://doi.org/10.1038/nri2515.
163. Powell N, Walker MM, Talley NJ. The mucosal immune system: master regulator of bidirectional gut-brain communications. *Nat Rev Gastroenterol Hepatol.* 2017;14(3):143–159. https://doi.org/10.1038/nrgastro.2016.191.
164. Macpherson AJ, Harris NL. Interactions between commensal intestinal bacteria and the immune system. *Nat Rev Immunol.* 2004;4(6):478–485. https://doi.org/10.1038/nri1373.
165. Didierlaurent A, Sirard J-C, Kraehenbuhl J-P, Neutra MR. How the gut senses its content. *Cell Microbiol.* 2002;4(2):61–72. https://doi.org/10.1046/j.1462-5822.2002.00177.x.
166. Negi S, Das DK, Pahari S, Nadeem S, Agrewala JN. Potential role of gut microbiota in induction and regulation of innate immune memory. *Front Immunol.* 2019;10:2441. https://doi.org/10.3389/fimmu.2019.02441.
167. Kumar H, Kawai T, Akira S. Pathogen recognition by the innate immune system. *Int Rev Immunol.* 2011;30(1):16–34. https://doi.org/10.3109/08830185.2010.529976.
168. Takeda K, Akira S. Microbial recognition by toll-like receptors. *J Dermatol Sci.* 2004;34(2):73–82. https://doi.org/10.1016/j.jdermsci.2003.10.002.
169. Mazmanian SK, Cui HL, Tzianabos AO, Kasper DL. An immunomodulatory molecule of symbiotic bacteria directs maturation of the host immune system. *Cell.* 2005;122(1):107–118. https://doi.org/10.1016/j.cell.2005.05.007.
170. Gordon HA. Morphological and physiological characterization of germfree life. *Ann N Y Acad Sci.* 1959;78(1):208–220. https://doi.org/10.1111/j.1749-6632.1959.tb53104.x.
171. Glaister JR. Factors affecting the lymphoid cells in the small intestinal epithelium of the mouse. *Int Arch Allergy Appl Immunol.* 1973;45(5):719–730. https://doi.org/10.1159/000231071.
172. Moreau MC, Ducluzeau R, Guy-Grand D; Muller MC. Increase in the population of duodenal immunoglobulin A plasmocytes in axenic mice associated with different living or dead bacterial strains of intestinal origin. *Infect Immun.* 1978;21(2):532–539. https://doi.org/10.1128/iai.21.2.532-539.1978.

173. Matsumoto S, Setoyama H, Umesaki Y. Differential induction of major histocompatibility complex molecules on mouse intestine by bacterial colonization. *Gastroenterology.* 1992;103(6):1777–1782. https://doi.org/10.1016/0016-5085(92)91434-6.
174. Mikkelsen HB, Garbarsch C, Tranum-Jensen J, Thuneberg L. Macrophages in the small intestinal muscularis externa of embryos, newborn and adult germ-free mice. *J Mol Histol.* 2004;35(4):377–387. https://doi.org/10.1023/B:HIJO.0000039840.86420.b7.
175. Sobko T, Reinders C, Norin E, Midtvedt T, Gustafsson LE, Lundberg JO. Gastrointestinal nitric oxide generation in germ-free and conventional rats. *Am J Physiol Gastrointest Liver Physiol.* 2004;287(5):G993–G997. https://doi.org/10.1152/ajpgi.00203.2004.
176. Umesaki Y, Okada Y, Matsumoto S, Imaoka A, Setoyama H. Segmented filamentous bacteria are indigenous intestinal bacteria that activate intraepithelial lymphocytes and induce MHC class II molecules and fucosyl asialo GM1 glycolipids on the small intestinal epithelial cells in the ex-germ-free mouse. *Microbiol Immunol.* 1995;39(8):555–562. https://doi.org/10.1111/j.1348-0421.1995.tb02242.x.
177. Abdel-Haq R, Schlachetzki JCM, Glass CK, Mazmanian SK. Microbiome–microglia connections via the gut–brain axis. *J Exp Med.* 2019;216(1):41–59. https://doi.org/10.1084/jem.20180794.
178. Saijo K, Glass CK. Microglial cell origin and phenotypes in health and disease. *Nat Rev Immunol.* 2011;11(11):775–787. https://doi.org/10.1038/nri3086.
179. Hickman SE, Kingery ND, Ohsumi TK, et al. The microglial sensome revealed by direct RNA sequencing. *Nat Neurosci.* 2013;16(12):1896–1905. https://doi.org/10.1038/nn.3554.
180. Tremblay M-È, Stevens B, Sierra A, Wake H, Bessis A, Nimmerjahn A. The role of microglia in the healthy brain. *J Neurosci.* 2011;31(45):16064–16069. https://doi.org/10.1523/JNEUROSCI.4158-11.2011.
181. Wake H, Wohleb ES. Neuron-microglia interactions in mental health disorders: "for better, and for worse.". *Front Immunol.* 2016;7:544. https://doi.org/10.3389/fimmu.2016.00544.
182. Yamasaki R, Lu H, Butovsky O, et al. Differential roles of microglia and monocytes in the inflamed central nervous system. *J Exp Med.* 2014;211(8):1533–1549. https://doi.org/10.1084/jem.20132477.
183. Ransohoff RM, El Khoury J. Microglia in health and disease. *Cold Spring Harb Perspect Biol.* 2016;8(1):a020560. https://doi.org/10.1101/cshperspect.a020560.
184. Wohleb ES. Neuron-microglia interactions in mental health disorders: "for better, and for worse.". *Front Immunol.* 2016;7(NOV):544. https://doi.org/10.3389/fimmu.2016.00544.
185. Wohleb ES, DB MK, Sheridan JF, Godbout JP. Monocyte trafficking to the brain with stress and inflammation: a novel axis of immune-to-brain communication that influences mood and behavior. *Front Neurosci.* 2015;9(JAN):447. https://doi.org/10.3389/fnins.2014.00447.
186. Schafer DP, Stevens B. Microglia function in central nervous system development and plasticity. *Cold Spring Harb Perspect Biol.* 2015;7(10):a020545. https://doi.org/10.1101/cshperspect.a020545.
187. Erny D, De Angelis ALH, Jaitin D, et al. Host microbiota constantly control maturation and function of microglia in the CNS. *Nat Neurosci.* 2015;18(7):965–977. https://doi.org/10.1038/nn.4030.

188. Thion MS, Low D, Silvin A, et al. Microbiome influences prenatal and adult microglia in a sex-specific manner. *Cell.* 2018;172(3):500–516.e16. https://doi.org/10.1016/j.cell.2017.11.042.
189. Shea-Donohue T, Urban JF. Neuroimmune modulation of gut function. *Handb Exp Pharmacol.* 2017;239:247–267. https://doi.org/10.1007/164_2016_109.
190. Kayama H, Takeda K. Regulation of intestinal homeostasis by innate and adaptive immunity. *Int Immunol.* 2012;24(11):673–680. https://doi.org/10.1093/intimm/dxs094.
191. Di Mauro A, Neu J, Riezzo G, et al. Gastrointestinal function development and microbiota. *Ital J Pediatr.* 2013;39(1):15. https://doi.org/10.1186/1824-7288-39-15.
192. Fujiwara D, Inoue S, Wakabayashi H, Fujii T. The anti-allergic effects of lactic acid Bacteria are strain dependent and mediated by effects on both Th1/Th2 cytokine expression and balance. *Int Arch Allergy Immunol.* 2004;135(3):205–215. https://doi.org/10.1159/000081305.
193. Gill HS, Rutherfurd KJ, Prasad J, Gopal PK. Enhancement of natural and acquired immunity by lactobacillus rhamnosus (HN001), Lactobacillus acidophilus (HN017) and Bifidobacterium lactis (HN019). *Br J Nutr.* 2000;83(2):167–176. https://doi.org/10.1017/s0007114500000210.
194. Akira S, Hemmi H. Recognition of pathogen-associated molecular patterns by TLR family. *Immunol Lett.* 2003;85(2):85–95. https://doi.org/10.1016/S0165-2478(02)00228-6.
195. Ulevitch RJ. Immunology: toll gates for pathogen selection. *Nature.* 1999;401(6755):755–756. https://doi.org/10.1038/44490.
196. O'Hara AM, Shanahan F. The gut flora as a forgotten organ. *EMBO Rep.* 2006;7(7):688–693. https://doi.org/10.1038/sj.embor.7400731.
197. Filiano AJ, Gadani SP, Kipnis J. Interactions of innate and adaptive immunity in brain development and function. *Brain Res.* 1617;2015:18–27. https://doi.org/10.1016/j.brainres.2014.07.050.
198. Amor S, Woodroofe MN. Innate and adaptive immune responses in neurodegeneration and repair. *Immunology.* 2014;141(3):287–291. https://doi.org/10.1111/imm.12134.
199. Brachman RA, Lehmann ML, Maric D, Herkenham M. Lymphocytes from chronically stressed mice confer antidepressant-like effects to naive mice. *J Neurosci.* 2015;35(4):1530–1538. https://doi.org/10.1523/JNEUROSCI.2278-14.2015.
200. *Shaping of Intestinal Microbiota in Nlrp6- and Rag2-Deficient Mice Depends on Community Structure.* PubMed - NCBI https://www.ncbi.nlm.nih.gov/pubmed/?term=10.1016%2Fj.celrep.2017.12.027. Accessed February 10, 2020.
201. Cervantes-Barragan L, Chai JN, Tianero MD, et al. Lactobacillus reuteri induces gut intraepithelial CD4+CD8αα+ T cells. *Science.* 2017;357(6353):806–810. https://doi.org/10.1126/science.aah5825.
202. Bharwani A, Mian MF, Foster JA, Surette MG, Bienenstock J, Forsythe P. Structural & functional consequences of chronic psychosocial stress on the microbiome & host. *Psychoneuroendocrinology.* 2016;63:217–227. https://doi.org/10.1016/j.psyneuen.2015.10.001.
203. Smith CJ, Emge JR, Berzins K, et al. Probiotics normalize the gut-brain-microbiota axis in immunodeficient mice. *Am J Physiol Gastrointest Liver Physiol.* 2014;307(8):G793–G802. https://doi.org/10.1152/ajpgi.00238.2014.
204. Wegorzewska MM, Glowacki RWP, Hsieh SA, et al. Diet modulates colonic T cell responses by regulating the expression of a bacteroides thetaiotaomicron antigen. *Sci Immunol.* 2019;4(32):eaau9079. https://doi.org/10.1126/sciimmunol.aau9079.
205. Furness JB, Callaghan BP, Rivera LR, Cho HJ. The Enteric Nervous System and Gastrointestinal Innervation: Integrated Local and Central Control. In: Lyte M,

Cryan J, eds. *Microbial Endocrinology: The Microbiota-Gut-Brain Axis in Health and Disease.* New York, NY: Springer; 2014. Adv Exp Med Biol; vol 817. https://doi.org/10.1007/978-1-4939-0897-4_3.
206. Furness JB. The enteric nervous system and neurogastroenterology. *Nat Rev Gastroenterol Hepatol.* 2012;9(5):286–294. https://doi.org/10.1038/nrgastro.2012.32.
207. Vermeulen W, De Man JG, Pelckmans PA, De Winter BY. Neuroanatomy of lower gastrointestinal pain disorders. *World J Gastroenterol.* 2014;20(4):1005–1020. https://doi.org/10.3748/wjg.v20.i4.1005.
208. Baj A, Bistoletti M, Bosi A, Moro E, Giaroni C, Crema F. Marine toxins and nociception: potential therapeutic use in the treatment of visceral pain associated with gastrointestinal disorders. *Toxins.* 2019;11(8):449. https://doi.org/10.3390/toxins11080449.
209. Ceccotti C, Giaroni C, Bistoletti M, Viola M, Crema F, Terova G. Neurochemical characterization of myenteric neurons in the juvenile gilthead sea bream (*Sparus aurata*) intestine. *PLoS One.* 2018;13(8):e0201760. https://doi.org/10.1371/journal.pone.0201760.
210. Clerc N, Furness JB. Intrinsic primary afferent neurones of the digestive tract. *Neurogastroenterol Motil.* 2004;16(suppl 1):24–27. https://doi.org/10.1111/j.1743-3150.2004.00470.x.
211. Giaroni C. Purinergic signalling and development of the autonomic nervous system. *Auton Neurosci.* 2015;191:67–77. https://doi.org/10.1016/j.autneu.2015.04.009.
212. Filpa V, Carpanese E, Marchet S, et al. Nitric oxide regulates homeoprotein OTX1 and OTX2 expression in the rat myenteric plexus after intestinal ischemia-reperfusion injury. *Am J Physiol Gastrointest Liver Physiol.* 2017;312(4):G374–G389. https://doi.org/10.1152/ajpgi.00386.2016.
213. Gulbransen BD, Sharkey KA. Novel functional roles for enteric glia in the gastrointestinal tract. *Nat Rev Gastroenterol Hepatol.* 2012;9(11):625–632. https://doi.org/10.1038/nrgastro.2012.138.
214. Bin A, Caputi V, Bistoletti M, et al. The ecto-enzymes CD73 and adenosine deaminase modulate 5′-AMP-derived adenosine in myofibroblasts of the rat small intestine. *Purinergic Signal.* 2018;14:409–421. https://doi.org/10.1007/s11302-018-9623-6.
215. Carpanese E, Moretto P, Filpa V, et al. Antagonism of ionotropic glutamate receptors attenuates chemical ischemia-induced injury in rat primary cultured myenteric ganglia. *PLoS One.* 2014;9(11):e113613. https://doi.org/10.1371/journal.pone.0113613.
216. Bistoletti M, Caputi V, Baranzini N, et al. Antibiotic treatment-induced dysbiosis differently affects BDNF and TrkB expression in the brain and in the gut of juvenile mice. *PLoS One.* 2019;14(2):e0212856. https://doi.org/10.1371/journal.pone.0212856.
217. Alverdy J, Holbrook C, Rocha F, et al. Gut-derived sepsis occurs when the right pathogen with the right virulence genes meets the right host: evidence for in vivo virulence expression in *Pseudomonas aeruginosa. Ann Surg.* 2000;232(4):480–489. https://doi.org/10.1097/00000658-200010000-00003.
218. Rolig AS, Mittge EK, Ganz J, et al. The enteric nervous system promotes intestinal health by constraining microbiota composition. *PLoS Biol.* 2017;15:e2000689. https://doi.org/10.1371/journal.pbio.2000689.
219. Hyland NP, Cryan JF. Microbe-host interactions: influence of the gut microbiota on the enteric nervous system. *Dev Biol.* 2016;417(2):182–187. https://doi.org/10.1016/j.ydbio.2016.06.027.
220. Collins J, Borojevic R, Verdu EF, Huizinga JD, Ratcliffe EM. Intestinal microbiota influence the early postnatal development of the enteric nervous system. *Neurogastroenterol Motil.* 2014;26(1):98–107. https://doi.org/10.1111/nmo.12236.

221. Mcvey Neufeld KA, Perez-Burgos A, Mao YK, Bienenstock J, Kunze WA. The gut microbiome restores intrinsic and extrinsic nerve function in germ-free mice accompanied by changes in calbindin. *Neurogastroenterol Motil.* 2015;27(5):627–636. https://doi.org/10.1111/nmo.12534.
222. Caputi V, Marsilio I, Filpa V, et al. Antibiotic-induced dysbiosis of the microbiota impairs gut neuromuscular function in juvenile mice. *Br J Pharmacol.* 2017;174 (20):3623–3639. https://doi.org/10.1111/bph.13965.
223. Kabouridis PS, Pachnis V. Emerging roles of gut microbiota and the immune system in the development of the enteric nervous system. *J Clin Investig.* 2015;125(3):956–964. https://doi.org/10.1172/JCI76308.
224. Anitha M, Vijay-Kumar M, Sitaraman SV, Gewirtz AT, Srinivasan S. Gut microbial products regulate murine gastrointestinal motility via toll-like receptor 4 signaling. *Gastroenterology.* 2012;143(4):1006. https://doi.org/10.1053/j.gastro.2012.06.034.
225. Caputi V, Marsilio I, Cerantola S, et al. Toll-like receptor 4 modulates small intestine neuromuscular function through nitrergic and purinergic pathways. *Front Pharmacol.* 2017;8(JUN):350. https://doi.org/10.3389/fphar.2017.00350.
226. Brun P, Giron MC, Qesari M, et al. Toll-like receptor 2 regulates intestinal inflammation by controlling integrity of the enteric nervous system. *Gastroenterology.* 2013;145 (6):1323–1333. https://doi.org/10.1053/j.gastro.2013.08.047.
227. Dass NB, John AK, Bassil AK, et al. The relationship between the effects of short-chain fatty acids on intestinal motility in vitro and GPR43 receptor activation. *Neurogastroenterol Motil.* 2007;19(1):66–74. https://doi.org/10.1111/j.1365-2982.2006.00853.x.
228. Husebye E, Hellström PM, Sundler F, Chen J, Midtvedt T. Influence of microbial species on small intestinal myoelectric activity and transit in germ-free rats. *Am J Physiol Gastrointest Liver Physiol.* 2001;280(3):G368–G380. https://doi.org/10.1152/ajpgi.2001.280.3.g368.
229. Mcvey Neufeld KA, Mao YK, Bienenstock J, Foster JA, Kunze WA. The microbiome is essential for normal gut intrinsic primary afferent neuron excitability in the mouse. *Neurogastroenterol Motil.* 2013;25(2):183–e88. https://doi.org/10.1111/nmo.12049.
230. Kunze WA, Mao Y-K, Wang B, et al. Lactobacillus reuteri enhances excitability of colonic AH neurons by inhibiting calcium-dependent potassium channel opening. *J Cell Mol Med.* 2009;13(8 B):2261–2270. https://doi.org/10.1111/j.1582-4934.2009.00686.x.
231. Mao Y-K, Kasper DL, Wang B, Forsythe P, Bienenstock J, Kunze WA. Bacteroides fragilis polysaccharide A is necessary and sufficient for acute activation of intestinal sensory neurons. *Nat Commun.* 2013;4. https://doi.org/10.1038/ncomms2478.
232. Al-Nedawi K, Mian MF, Hossain N, et al. Gut commensal microvesicles reproduce parent bacterial signals to host immune and enteric nervous systems. *FASEB J.* 2015; 29(2):684–695. https://doi.org/10.1096/fj.14-259721.
233. Carabotti M, Scirocco A, Maselli MA, Severi C. The gut-brain axis: Interactions between enteric microbiota, central and enteric nervous systems. *Ann Gastroenterol.* 2015;28(2):203–209.
234. Mulak A, Bonaz B. Irritable bowel syndrome: a model of the brain-gut interactions. *Med Sci Monit.* 2004;10(4):RA55–RA62.
235. O'Mahony SM, Hyland NP, Dinan TG, Cryan JF. Maternal separation as a model of brain-gut axis dysfunction. *Psychopharmacology (Berl).* 2011;214(1):71–88. https://doi.org/10.1007/s00213-010-2010-9.
236. Jones MP, Dilley JB, Drossman D, Crowell MD. Brain-gut connections in functional GI disorders: anatomic and physiologic relationships. *Neurogastroenterol Motil.* 2006;18 (2):91–103. https://doi.org/10.1111/j.1365-2982.2005.00730.x.
237. Almeida TF, Roizenblatt S, Tufik S. Afferent pain pathways: a neuroanatomical review. *Brain Res.* 2004;1000(1–2):40–56. https://doi.org/10.1016/j.brainres.2003.10.073.

238. Yuan H, Silberstein SD. Vagus nerve and vagus nerve stimulation, a comprehensive review: part II. *Headache*. 2016;56(2):259–266. https://doi.org/10.1111/head.12650.
239. Grundy D. Neuroanatomy of visceral nociception: vagal and splanchnic afferent. *Gut*. 2002;51(suppl 1):i2–i5. https://doi.org/10.1136/gut.51.suppl_1.i2.
240. Page A, Young R, Martin C, Umaerus M. Metabotropic glutamate receptors inhibit mechanosensitivity in vagal sensory neurons. *Gastroenterology*. 2005;128(2):402–410. https://doi.org/10.1053/j.gastro.2004.11.062.
241. Schonhoff SE, Giel-Moloney M, Leiter AB. Minireview: development and differentiation of gut endocrine cells. *Endocrinology*. 2004;145(6):2639–2644. https://doi.org/10.1210/en.2004-0051.
242. O'Hara JR, Ho W, Linden DR, Mawe GM, Sharkey KA. Enteroendocrine cells and 5-HT availability are altered in mucosa of guinea pigs with TNBS ileitis. *Am J Physiol Gastrointest Liver Physiol*. 2004;287(5):G998–G1007. https://doi.org/10.1152/ajpgi.00090.2004.
243. Raybould HE, Cooke HJ, Christofi FL. Sensory mechanisms: transmitters, modulators and reflexes. *Neurogastroenterol Motil*. 2004;16(s1):60–63. https://doi.org/10.1111/j.1743-3150.2004.00477.x.
244. Holzer P. Afferent signalling of gastric acid challenge. *J Physiol Pharmacol*. 2003;54(suppl 4):43–53.
245. Goehler LE, Gaykema RPA, Opitz N, Reddaway R, Badr N, Lyte M. Activation in vagal afferents and central autonomic pathways: early responses to intestinal infection with Campylobacter jejuni. *Brain Behav Immun*. 2005;19(4):334–344. https://doi.org/10.1016/j.bbi.2004.09.002.
246. Lyte M, Varcoe JJ, Bailey MT. Anxiogenic effect of subclinical bacterial infection in mice in the absence of overt immune activation. *Physiol Behav*. 1998;65(1):63–68. https://doi.org/10.1016/S0031-9384(98)00145-0.
247. Klarer M, Krieger JP, Richetto J, et al. Abdominal vagal afferents modulate the brain transcriptome and behaviors relevant to schizophrenia. *J Neurosci*. 2018;38(7):1634–1647. https://doi.org/10.1523/JNEUROSCI.0813-17.2017.
248. Desbonnet L, Clarke G, Shanahan F, Dinan TG, Cryan JF. Microbiota is essential for social development in the mouse. *Mol Psychiatry*. 2014;19(2):146–148. https://doi.org/10.1038/mp.2013.65.
249. Perez-Burgos A, Wang B, Mao Y-K, et al. Psychoactive bacteria Lactobacillus rhamnosus (JB-1) elicits rapid frequency facilitation in vagal afferents. *Am J Physiol Gastrointest Liver Physiol*. 2013;304(2):G211–G220. https://doi.org/10.1152/ajpgi.00128.2012.
250. Bravo JA, Forsythe P, Chew MV, et al. Ingestion of Lactobacillus strain regulates emotional behavior and central GABA receptor expression in a mouse via the vagus nerve. *Proc Natl Acad Sci U S A*. 2011;108(38):16050–16055. https://doi.org/10.1073/pnas.1102999108.
251. Sgritta M, Dooling SW, Buffington SA, et al. Mechanisms underlying microbial-mediated changes in social behavior in mouse models of autism spectrum disorder. *Neuron*. 2019;101(2):246–259.e6. https://doi.org/10.1016/j.neuron.2018.11.018.
252. Vazquez E, Barranco A, Ramirez M, et al. Dietary 2'-fucosyllactose enhances operant conditioning and long-term potentiation via gut-brain communication through the vagus nerve in rodents. Scavone C, Ed. *PLoS One*. 2016;11(11):e0166070. https://doi.org/10.1371/journal.pone.0166070.
253. Bercik P, Denou E, Collins J, et al. The intestinal microbiota affect central levels of brain-derived neurotropic factor and behavior in mice. *Gastroenterology*. 2011;141(2):599–609.e3. https://doi.org/10.1053/j.gastro.2011.04.052.

254. Van Felius ID, LMA A, Bosscha K, et al. Interdigestive small bowel motility and duodenal bacterial overgrowth in experimental acute pancreatitis. *Neurogastroenterol Motil.* 2003;15(3):267–276. https://doi.org/10.1046/j.1365-2982.2003.00410.x.
255. Maier TV, Lucio M, Lee LH, et al. Impact of dietary resistant starch on the human gut microbiome, metaproteome, and metabolome. Moran MA, ed. *MBio.* 2017;8(5): e01343-17. https://doi.org/10.1128/mBio.01343-17.
256. Sudo N, Chida Y, Aiba Y, et al. Postnatal microbial colonization programs the hypothalamic-pituitary-adrenal system for stress response in mice. *J Physiol.* 2004; 558(pt 1):263–275. https://doi.org/10.1113/jphysiol.2004.063388.
257. Clarke G, Grenham S, Scully P, et al. The microbiome-gut-brain axis during early life regulates the hippocampal serotonergic system in a sex-dependent manner. *Mol Psychiatry.* 2013;18(6):666–673. https://doi.org/10.1038/mp.2012.77.
258. Maqsood R, Stone TW. The gut-brain axis, BDNF, NMDA and CNS disorders. *Neurochem Res.* 2016;41(11):2819–2835. https://doi.org/10.1007/s11064-016-2039-1.
259. Klug M, Hill RA, Choy KHC, Kyrios M, Hannan AJ, van den Buuse M. Long-term behavioral and NMDA receptor effects of young-adult corticosterone treatment in BDNF heterozygous mice. *Neurobiol Dis.* 2012;46(3):722–731. https://doi.org/10.1016/j.nbd.2012.03.015.
260. Suen P-C, Wu K, Levine ES, et al. Brain-derived neurotrophic factor rapidly enhances phosphorylation of the postsynaptic N-methyl-D-aspartate receptor subunit 1. *Proc Natl Acad Sci U S A.* 1997;94(15):8191–8195. https://doi.org/10.1073/pnas.94.15.8191.
261. Moloney RD, Johnson AC, O'Mahony SM, Dinan TG, Greenwood-Van Meerveld B, Cryan JF. Stress and the microbiota-gut-brain axis in visceral PaRelevance to irritable bowel syndrome. *CNS Neurosci Ther.* 2016;22(2):102–117. https://doi.org/10.1111/cns.12490.
262. Lee C, Doo E, Choi JM, et al. The increased level of depression and anxiety in irritable bowel syndrome patients compared with healthy controls: systematic review and meta-analysis. *J Neurogastroenterol Motil.* 2017;23(3):349–362. https://doi.org/10.5056/jnm16220.
263. Eisenstein M. Microbiome: bacterial broadband. *Nature.* 2016;533(7603):S104–S106. https://doi.org/10.1038/533S104a.
264. Dinan TG, Quigley EMM, Ahmed SMM, et al. Hypothalamic-pituitary-gut axis dysregulation in irritable bowel syndrome: plasma cytokines as a potential biomarker? *Gastroenterology.* 2006;130(2):304–311. https://doi.org/10.1053/j.gastro.2005.11.033.
265. Hosoi T, Okuma Y, Nomura Y. Electrical stimulation of afferent vagus nerve induces IL-1β expression in the brain and activates HPA axis. *Am J Physiol Regul Integr Comp Physiol.* 2000;279(1):R141–R147. https://doi.org/10.1152/ajpregu.2000.279.1.r141.
266. Iyer LM, Aravind L, Coon SL, Klein DC, Koonin EV. Evolution of cell-cell signaling in animals: did late horizontal gene transfer from bacteria have a role? *Trends Genet.* 2004;20(7):292–299. https://doi.org/10.1016/j.tig.2004.05.007.
267. Lyte M, Ernst S. Alpha and beta adrenergic receptor involvement in catecholamine-induced growth of gram-negative bacteria. *Biochem Biophys Res Commun.* 1993; 190(2):447–452. https://doi.org/10.1006/bbrc.1993.1068.
268. Lyte M, Ernst S. Catecholamine induced growth of gram negative bacteria. *Life Sci.* 1992;50(3):203–212. https://doi.org/10.1016/0024-3205(92)90273-R.
269. Kennedy PJ, Cryan JF, Dinan TG, Clarke G. Kynurenine pathway metabolism and the microbiota-gut-brain axis. *Neuropharmacology.* 2017;112(pt B):399–412. https://doi.org/10.1016/j.neuropharm.2016.07.002.
270. Dehhaghi M, Kazemi Shariat Panahi H, Guillemin GJ. Microorganisms, tryptophan metabolism, and kynurenine pathway: a complex interconnected loop influencing human health status. *Int J Tryptophan Res.* 2019;12. https://doi.org/10.1177/1178646919852996.

271. Waclawiková B, El Aidy S. Role of microbiota and tryptophan metabolites in the remote effect of intestinal inflammation on brain and depression. *Pharmaceuticals*. 2018;11(3):63. https://doi.org/10.3390/ph11030063.
272. Gershon MD. 5-Hydroxytryptamine (serotonin) in the gastrointestinal tract. *Curr Opin Endocrinol Diabetes Obes*. 2013;20(1):14–21. https://doi.org/10.1097/MED.0b013e32835bc703.
273. Badawy AA-B. Tryptophan availability for kynurenine pathway metabolism across the life span: control mechanisms and focus on aging, exercise, diet and nutritional supplements. *Neuropharmacology*. 2017;112(pt B):248–263. https://doi.org/10.1016/j.neuropharm.2015.11.015.
274. Richard DM, Dawes MA, Mathias CW, Acheson A, Hill-Kapturczak N, Dougherty DM. L-tryptophan: basic metabolic functions, behavioral research and therapeutic indications. *Int J Tryptophan Res*. 2009;2:45–60. https://doi.org/10.4137/ijtr.s2129.
275. Yano JM, Yu K, Donaldson GP, et al. Indigenous bacteria from the gut microbiota regulate host serotonin biosynthesis. *Cell*. 2015;161(2):264–276. https://doi.org/10.1016/j.cell.2015.02.047.
276. Alemi F, Poole DP, Chiu J, et al. The receptor TGR5 mediates the prokinetic actions of intestinal bile acids and is required for normal defecation in mice. *Gastroenterology*. 2013;144(1):145–154. https://doi.org/10.1053/j.gastro.2012.09.055.
277. Reigstad CS, Salmonson CE, Rainey JF, et al. Gut microbes promote colonic serotonin production through an effect of short-chain fatty acids on enterochromaffin cells. *FASEB J*. 2015;29(4):1395–1403. https://doi.org/10.1096/fj.14-259598.
278. Kidd M, Gustafsson BI, Drozdov I, Modlin IM. IL1β- and LPS-induced serotonin secretion is increased in EC cells derived from Crohn's disease. *Neurogastroenterol Motil*. 2009;21(4):439–450. https://doi.org/10.1111/j.1365-2982.2008.01210.x.
279. Hata T, Asano Y, Yoshihara K, et al. Regulation of gut luminal serotonin by commensal microbiota in mice. *PLoS One*. 2017;12(7):e0180745. https://doi.org/10.1371/journal.pone.0180745.
280. Cao Y-N, Feng L-J, Liu Y-Y, et al. Effect of Lactobacillus rhamnosus GG supernatant on serotonin transporter expression in rats with post-infectious irritable bowel syndrome. *World J Gastroenterol*. 2018;24:338–350. https://doi.org/10.3748/wjg.v24.i3.338.
281. De Vadder F, Grasset E, Holm LM, et al. Gut microbiota regulates maturation of the adult enteric nervous system via enteric serotonin networks. *Proc Natl Acad Sci U S A*. 2018;115:6458–6463. https://doi.org/10.1073/pnas.1720017115.
282. El-Merahbi R, Löffler M, Mayer A, Sumara G. The roles of peripheral serotonin in metabolic homeostasis. *FEBS Lett*. 2015;589(15):1728–1734. https://doi.org/10.1016/j.febslet.2015.05.054.
283. Golubeva AV, Joyce SA, Moloney G, et al. Microbiota-related changes in bile acid & tryptophan metabolism are associated with gastrointestinal dysfunction in a mouse model of autism. *EBioMedicine*. 2017;24:166–178. LK. https://doi.org/10.1016/j.ebiom.2017.09.020.
284. Zhong W, Hutchinson TE, Chebolu S, Darmani NA. Serotonin 5-HT3 receptor-mediated vomiting occurs via the activation of Ca2+/CaMKII-dependent ERK1/2 signaling in the least shrew (*Cryptotis parva*). *PLoS One*. 2014;9(8):e104718. https://doi.org/10.1371/journal.pone.0104718.
285. Cussotto S, Sandhu KV, Dinan TG, Cryan JF. The neuroendocrinology of the microbiota-gut-brain axis: a behavioural perspective. *Front Neuroendocrinol*. 2018;51:80–101. https://doi.org/10.1016/j.yfrne.2018.04.002.
286. Foley S, Garsed K, Singh G, et al. Impaired uptake of serotonin by platelets from patients with irritable bowel syndrome correlates with duodenal immune activation. *Gastroenterology*. 2011;140(5):1434-43.e1. https://doi.org/10.1053/j.gastro.2011.01.052.

287. Kerckhoffs APM, ter Linde JJM, Akkermans LMA, Samsom M. SERT and TPH-1 mRNA expression are reduced in irritable bowel syndrome patients regardless of visceral sensitivity state in large intestine. *Am J Physiol Gastrointest Liver Physiol.* 2012;302(9):G1053–G1060. https://doi.org/10.1152/ajpgi.00153.2011.
288. Camilleri M. Serotonergic modulation of visceral sensation: lower gut. *Gut.* 2002;51: i81–i86. https://doi.org/10.1136/gut.51.suppl_1.i81.
289. Pusceddu MM, Gareau MG. Visceral pain: gut microbiota, a new hope? *J Biomed Sci.* 2018;25:73. https://doi.org/10.1186/s12929-018-0476-7.
290. Biaggini K, Barbey C, Borrel V, Feuilloley M, Déchelotte P, Connil N. The pathogenic potential of *Pseudomonas fluorescens* MFN1032 on enterocytes can be modulated by serotonin, substance P and epinephrine. *Arch Microbiol.* 2015;197(8):983–990. https://doi.org/10.1007/s00203-015-1135-y.
291. Agus A, Planchais J, Sokol H. Gut microbiota regulation of tryptophan metabolism in health and disease. *Cell Host Microbe.* 2018;23(6):716–724. https://doi.org/10.1016/j.chom.2018.05.003.
292. Ciorba MA. Indoleamine 2,3 dioxygenase in intestinal disease. *Curr Opin Gastroenterol.* 2013;29(2):146–152. https://doi.org/10.1097/MOG.0b013e32835c9cb3.
293. Schwarcz R, Bruno JP, Muchowski PJ, Wu HQ. Kynurenines in the mammalian brain: when physiology meets pathology. *Nat Rev Neurosci.* 2012;13(7):465–477. https://doi.org/10.1038/nrn3257.
294. Lugo-Huitrón R, Ugalde Muñiz P, Pineda B, Pedraza-Chaverrí J, Ríos C, Pérez-De La Cruz V. Quinolinic acid: an endogenous neurotoxin with multiple targets. *Oxid Med Cell Longev.* 2013;2013:104024. https://doi.org/10.1155/2013/104024.
295. Filpa V, Moro E, Protasoni M, Crema F, Frigo G, Giaroni C. Role of glutamatergic neurotransmission in the enteric nervous system and brain-gut axis in health and disease. *Neuropharmacology.* 2016;111:14–33. https://doi.org/10.1016/j.neuropharm.2016.08.024.
296. Filpa V, Carpanese E, Marchet S, et al. Interaction between NMDA glutamatergic and nitrergic enteric pathways during in vitro ischemia and reperfusion. *Eur J Pharmacol.* 2015;750:123–131. https://doi.org/10.1016/j.ejphar.2015.01.021.
297. Sanchez S, Rodríguez-Sanoja R, Ramos A, Demain AL. Our microbes not only produce antibiotics, they also overproduce amino acids. *J Antibiot.* 2018;71(1):26–36. https://doi.org/10.1038/ja.2017.142.
298. Forrest CM, McNair K, Pisar M, Khalil OS, Darlington LG, Stone TW. Altered hippocampal plasticity by prenatal kynurenine administration, kynurenine-3-monoxygenase (KMO) deletion or galantamine. *Neuroscience.* 2015;310:91–105. https://doi.org/10.1016/j.neuroscience.2015.09.022.
299. Khalil OS, Pisar M, Forrest CM, Vincenten MCJ, Darlington LG, Stone TW. Prenatal inhibition of the kynurenine pathway leads to structural changes in the hippocampus of adult rat offspring. *Eur J Neurosci.* 2014;39(10):1558–1571. https://doi.org/10.1111/ejn.12535.
300. Pisar M, Forrest CM, Khalil OS, et al. Modified neocortical and cerebellar protein expression and morphology in adult rats following prenatal inhibition of the kynurenine pathway. *Brain Res.* 2014;1576:1–17. https://doi.org/10.1016/j.brainres.2014.06.016.
301. O'Farrell K, Harkin A. Stress-related regulation of the kynurenine pathway: relevance to neuropsychiatric and degenerative disorders. *Neuropharmacology.* 2017;112(pt B):307–323. https://doi.org/10.1016/j.neuropharm.2015.12.004.
302. Strasser B, Becker K, Fuchs D, Gostner JM. Kynurenine pathway metabolism and immune activation: peripheral measurements in psychiatric and co-morbid conditions. *Neuropharmacology.* 2017;112(pt B):286–296. https://doi.org/10.1016/j.neuropharm.2016.02.030.

303. Mardinoglu A, Shoaie S, Bergentall M, et al. The gut microbiota modulates host amino acid and glutathione metabolism in mice. *Mol Syst Biol.* 2015;11(10):834. https://doi.org/10.15252/msb.20156487.
304. Cryan JF, Dinan TG. Mind-altering microorganisms: the impact of the gut microbiota on brain and behaviour. *Nat Rev Neurosci.* 2012;13(10):701–712. https://doi.org/10.1038/nrn3346.
305. Wikoff WR, Anfora AT, Liu J, et al. Metabolomics analysis reveals large effects of gut microflora on mammalian blood metabolites. *Proc Natl Acad Sci U S A.* 2009;106(10):3698–3703. https://doi.org/10.1073/pnas.0812874106.
306. Desbonnet L, Garrett L, Clarke G, Bienenstock J, Dinan TG. The probiotic Bifidobacteria infantis: an assessment of potential antidepressant properties in the rat. *J Psychiatr Res.* 2008;43(2):164–174. https://doi.org/10.1016/j.jpsychires.2008.03.009.
307. >Moloney GM, O'Leary OF, Salvo-Romero E, et al. Microbial regulation of hippocampal miRNA expression: implications for transcription of kynurenine pathway enzymes. *Behav Brain Res.* 2017;334:50–54. https://doi.org/10.1016/j.bbr.2017.07.026.
308. Martin-Gallausiaux C, Larraufie P, Jarry A, et al. Butyrate produced by commensal bacteria down-regulates indolamine 2,3-dioxygenase 1 (IDO-1) expression via a dual mechanism in human intestinal epithelial cells. *Front Immunol.* 2018;9:2838. https://doi.org/10.3389/fimmu.2018.02838.
309. Freewan M, Rees MD, Plaza TSS, et al. Human indoleamine 2,3-dioxygenase is a catalyst of physiological heme peroxidase reactions: implications for the inhibition of dioxygenase activity by hydrogen peroxide. *J Biol Chem.* 2013;288(3):1548–1567. https://doi.org/10.1074/jbc.M112.410993.
310. Valladares R, Bojilova L, Potts AH, et al. Lactobacillus johnsonii inhibits indoleamine 2,3-dioxygenase and alters tryptophan metabolite levels in BioBreeding rats. *FASEB J.* 2013;27(4):1711–1720. https://doi.org/10.1096/fj.12-223339.
311. Marin IA, Goertz JE, Ren T, et al. Microbiota alteration is associated with the development of stress-induced despair behavior. *Sci Rep.* 2017;7:43859. https://doi.org/10.1038/srep43859.
312. Wichers MC, Koek GH, Robaeys G, Verkerk R, Scharpé S, Maes M. IDO and interferon-alpha-induced depressive symptoms: a shift in hypothesis from tryptophan depletion to neurotoxicity. *Mol Psychiatry.* 2005;10(6):538–544. https://doi.org/10.1038/sj.mp.4001600.
313. Kelly JR, Borre Y, O' Brien C, et al. Transferring the blues: depression-associated gut microbiota induces neurobehavioural changes in the rat. *J Psychiatr Res.* 2016;82:109–118. https://doi.org/10.1016/j.jpsychires.2016.07.019.
314. Terbeck S, Savulescu J, Chesterman LP, Cowen PJ. Noradrenaline effects on social behaviour, intergroup relations, and moral decisions. *Neurosci Biobehav Rev.* 2016;66:54–60. https://doi.org/10.1016/j.neubiorev.2016.03.031.
315. Szymańska K, Kuśmierska K, Demkow U. Inherited disorders of brain neurotransmitters: pathogenesis and diagnostic approach. *Adv Exp Med Biol.* 2015;837:1–8. https://doi.org/10.1007/5584_2014_86.
316. De Ponti F, Giaroni C, Cosentino M, Lecchini S, Frigo G. Adrenergic mechanisms in the control of gastrointestinal motility: from basic science to clinical applications. *Pharmacol Ther.* 1996;69(1):59–78. https://doi.org/10.1016/0163-7258(95)02031-4.
317. Costa M, Brookes SJH, Hennig GW. Anatomy and physiology of the enteric nervous system. *Gut.* 2000;47:iv15–iv19. https://doi.org/10.1136/gut.47.suppl_4.iv15.
318. Lyte M, Chapel A, Lyte JM, et al. Resistant starch alters the microbiota-gut brain axis: implications for dietary modulation of behavior. *PLoS One.* 2016;11(1):e0146406. https://doi.org/10.1371/journal.pone.0146406.

319. Cogan TA, Thomas AO, Rees LEN, et al. Norepinephrine increases the pathogenic potential of campylobacter jejuni. *Gut*. 2007;56(8):1060–1065. https://doi.org/10.1136/gut.2006.114926.
320. Sandrini SM, Shergill R, Woodward J, et al. Elucidation of the mechanism by which catecholamine stress hormones liberate iron from the innate immune defense proteins transferrin and lactoferrin. *J Bacteriol*. 2010;192(2):587–594. https://doi.org/10.1128/JB.01028-09.
321. Freestone PPE, Haigh RD, Williams PH, Lyte M. Involvement of enterobactin in norepinephrine-mediated iron supply from transferrin to enterohaemorrhagic *Escherichia coli*. *FEMS Microbiol Lett*. 2003;222(1):39–43. https://doi.org/10.1016/S0378-1097(03)00243-X.
322. Sandrini S, Aldriwesh M, Alruways M, Freestone P. Microbial endocrinology: host-bacteria communication within the gut microbiome. *J Endocrinol*. 2015;225(2): R21–R34. https://doi.org/10.1530/JOE-14-0615.
323. Lyte M, Villageliú DN, Crooker BA, Brown DR. Symposium review: microbial endocrinology-why the integration of microbes, epithelial cells, and neurochemical signals in the digestive tract matters to ruminant health. *J Dairy Sci*. 2018;101 (6):5619–5628. https://doi.org/10.3168/jds.2017-13589.
324. Green BT, Lyte M, Kulkarni-Narla A, Brown DR. Neuromodulation of enteropathogen internalization in Peyer's patches from porcine jejunum. *J Neuroimmunol*. 2003;141 (1–2):74–82. https://doi.org/10.1016/S0165-5728(03)00225-X.
325. Chen C, Lyte M, Stevens MP, Vulchanova L, Brown DR. Mucosally-directed adrenergic nerves and sympathomimetic drugs enhance non-intimate adherence of Escherichia coli O157:H7 to porcine cecum and colon. *Eur J Pharmacol*. 2006;539 (1–2):116–124. https://doi.org/10.1016/j.ejphar.2006.03.081.
326. Shishov VA, Kirovskaya TA, Kudrin VS, Oleskin AV. Amine neuromediators, their precursors, and oxidation products in the culture of *Escherichia coli* k-12. *Appl Biochem Microbiol*. 2009;45(5):494–497. https://doi.org/10.1134/S0003683809050068.
327. Tsavkelova EA, Botvinko IV, Kudrin VS, Oleskin AV. Detection of neurotransmitter amines in microorganisms with the use of high-performance liquid chromatography. *Dokl Biochem*. 2000;372(1–6):115–117.
328. Asano Y, Hiramoto T, Nishino R, et al. Critical role of gut microbiota in the production of biologically active, free catecholamines in the gut lumen of mice. *Am J Physiol Gastrointest Liver Physiol*. 2012;303:1288–1295. https://doi.org/10.1152/ajpgi.00341.2012.-There.
329. Lyte M, Brown DR. Evidence for PMAT- and OCT-like biogenic amine transporters in a probiotic strain of lactobacillus: Implications for interkingdom communication within the microbiota-gut-brain axis. *PLoS One*. 2018;13(1):e0191037. https://doi.org/10.1371/journal.pone.0191037.
330. Desbonnet L, Clarke G, Traplin A, et al. Gut microbiota depletion from early adolescence in mice: implications for brain and behaviour. *Brain Behav Immun*. 2015;48:165–173. https://doi.org/10.1016/j.bbi.2015.04.004.
331. Heijtz RD, Wang S, Anuar F, et al. Normal gut microbiota modulates brain development and behavior. *Proc Natl Acad Sci U S A*. 2011;108(7):3047–3052. https://doi.org/10.1073/pnas.1010529108.
332. Schmidt MJ, Mirnics K. Neurodevelopment, GABA system dysfunction, and schizophrenia. *Neuropsychopharmacology*. 2015;40(1):190–206. https://doi.org/10.1038/npp.2014.95.
333. Pessione E. Lactic acid bacteria contribution to gut microbiota complexity: lights and shadows. *Front Cell Infect Microbiol*. 2012;2:86. https://doi.org/10.3389/fcimb.2012.00086.

334. Tsai MF, McCarthy P, Miller C. Substrate selectivity in glutamate-dependent acid resistance in enteric bacteria. *Proc Natl Acad Sci U S A*. 2013;110(15):5898–5902. https://doi.org/10.1073/pnas.1301444110.
335. Matsumoto M, Kibe R, Ooga T, et al. Impact of intestinal microbiota on intestinal luminal metabolome. *Sci Rep*. 2012;2:233. https://doi.org/10.1038/srep00233.
336. Richard HT, Foster JW. Acid resistance in *Escherichia coli*. *Adv Appl Microbiol*. 2003;52:167–186. https://doi.org/10.1016/S0065-2164(03)01007-4.
337. Han TC, Kwon DH, Hegazy M, Lu CD. Transcriptome analysis of agmatine and putrescine catabolism in Pseudomonas aeruginosa PAO1. *J Bacteriol*. 2008;190 (6):1966–1975. https://doi.org/10.1128/JB.01804-07.
338. Barrett E, Ross RP, O'Toole PW, Fitzgerald GF, Stanton C. γ-Aminobutyric acid production by culturable bacteria from the human intestine. *J Appl Microbiol*. 2012;113 (2):411–417. https://doi.org/10.1111/j.1365-2672.2012.05344.x.
339. Lyte M, Vulchanova L, Brown DR. Stress at the intestinal surface: catecholamines and mucosa-bacteria interactions. *Cell Tissue Res*. 2011;343(1):23–32. https://doi.org/10.1007/s00441-010-1050-0.
340. Mazzoli R, Pessione E. The neuro-endocrinological role of microbial glutamate and GABA signaling. *Front Microbiol*. 2016;7(NOV):1934. https://doi.org/10.3389/fmicb.2016.01934.
341. Pokusaeva K, Johnson C, Luk B, et al. GABA-producing bifidobacterium dentium modulates visceral sensitivity in the intestine. *Neurogastroenterol Motil*. 2017;29(1): e12904. https://doi.org/10.1111/nmo.12904.
342. Cui Y, Miao K, Niyaphorn S, Qu X. Production of gamma-aminobutyric acid from lactic acid bacteria: a systematic review. *Int J Mol Sci*. 2020;21(3):995. https://doi.org/10.3390/ijms21030995.
343. Janik R, Thomason LAM, Stanisz AM, Forsythe P, Bienenstock J, Stanisz GJ. Magnetic resonance spectroscopy reveals oral Lactobacillus promotion of increases in brain GABA, N-acetyl aspartate and glutamate. *Neuroimage*. 2016;125:988–995. https://doi.org/10.1016/j.neuroimage.2015.11.018.
344. Liang L, Zhou H, Zhang S, Yuan J, Wu H. Effects of gut microbiota disturbance induced in early life on the expression of extrasynaptic GABA-A receptor α5 and δ subunits in the hippocampus of adult rats. *Brain Res Bull*. 2017;135:113–119. https://doi.org/10.1016/j.brainresbull.2017.09.014.
345. Burokas A, Arboleya S, Moloney RD, et al. Targeting the microbiota-gut-brain axis: prebiotics have anxiolytic and antidepressant-like effects and reverse the impact of chronic stress in mice. *Biol Psychiatry*. 2017;82(7):472–487. https://doi.org/10.1016/j.biopsych.2016.12.031.
346. Hyland NP, Cryan JF. A gut feeling about GABA: focus on GABAB receptors. *Front Pharmacol*. 2010;1:124. https://doi.org/10.3389/fphar.2010.00124.
347. Pérez-Berezo T, Pujo J, Martin P, et al. Identification of an analgesic lipopeptide produced by the probiotic *Escherichia coli* strain Nissle 1917. *Nat Commun*. 2017;8(1):1314. https://doi.org/10.1038/s41467-017-01403-9.
348. Auteri M, Zizzo MG, Serio R. GABA and GABA receptors in the gastrointestinal tract: from motility to inflammation. *Pharmacol Res*. 2015;93:11–21. https://doi.org/10.1016/j.phrs.2014.12.001.
349. Bjurstöm H, Wang JY, Ericsson I, et al. GABA, a natural immunomodulator of T lymphocytes. *J Neuroimmunol*. 2008;205(1–2):44–50. https://doi.org/10.1016/j.jneuroim.2008.08.017.
350. Guthrie GD, Nicholson-Guthrie CS. Gamma-aminobutyric acid uptake by a bacterial system with neurotransmitter binding characteristics. *Proc Natl Acad Sci U S A*. 1989;86 (19):7378–7381. https://doi.org/10.1073/pnas.86.19.7378.

351. Dagorn A, Hillion M, Chapalain A, et al. Gamma-aminobutyric acid acts as a specific virulence regulator in *Pseudomonas aeruginosa. Microbiology.* 2013;159(pt 2):339–351. https://doi.org/10.1099/mic.0.061267-0.
352. Briguglio M, Dell'Osso B, Panzica G, et al. Dietary neurotransmitters: a narrative review on current knowledge. *Nutrients.* 2018;10(5):591. https://doi.org/10.3390/nu10050591.
353. Janeczko MJ, Stoll B, Chang X, Guan X, Burrin DG. Extensive gut metabolism limits the intestinal absorption of excessive supplemental dietary glutamate loads in infant pigs. *J Nutr.* 2007;137(11):2384–2390.
354. Haÿs SP, Ordonez JM, Burrin DG, Sunehag AL. Dietary glutamate is almost entirely removed in its first pass through the splanchnic bed in premature infants. *Pediatr Res.* 2007;62(3):353–356. https://doi.org/10.1203/PDR.0b013e318123f719.
355. Nakayama Y, Hashimoto K-I, Sawada Y, Sokabe M, Kawasaki H, Martinac B. Corynebacterium glutamicum mechanosensitive channels: towards unpuzzling "glutamate efflux" for amino acid production. *Biophys Rev.* 2018;10(5):1359–1369. https://doi.org/10.1007/s12551-018-0452-1.
356. Zareian M, Ebrahimpour A, Bakar FA, et al. A glutamic acid-producing lactic acid bacteria isolated from malaysian fermented foods. *Int J Mol Sci.* 2012;13(5):5482–5497. https://doi.org/10.3390/ijms13055482.
357. Ger M-F, Rendon G, Tilson JL, Jakobsson E. Domain-based identification and analysis of glutamate receptor ion channels and their relatives in prokaryotes. *PLoS One.* 2010;5(10):e12827. https://doi.org/10.1371/journal.pone.0012827.
358. Genoux D, Montgomery JM. Glutamate receptor plasticity at excitatory synapses in the brain. *Clin Exp Pharmacol Physiol.* 2007;34(10):1058–1063. https://doi.org/10.1111/j.1440-1681.2007.04722.x.
359. Giaroni C, Marchet S, Carpanese E, et al. Role of neuronal and inducible nitric oxide synthases in the guinea pig ileum myenteric plexus during in vitro ischemia and reperfusion. *Neurogastroenterol Motil.* 2013;25(2):e114–e126. https://doi.org/10.1111/nmo.12061.
360. Gill S, Pulido O. Glutamate receptors in peripheral tissues: current knowledge, future research, and implications for toxicology. *Toxicol Pathol.* 2001;29:208–223. https://doi.org/10.1080/019262301317052486.
361. Giaroni C, Zanetti E, Chiaravalli AM, et al. Evidence for a glutamatergic modulation of the cholinergic function in the human enteric nervous system via NMDA receptors. *Eur J Pharmacol.* 2003;476(1–2):63–69. https://doi.org/10.1016/S0014-2999(03)02147-2.
362. Giuliani D, Giaroni C, Zanetti E, et al. Involvement of glutamate receptors of the NMDA type in the modulation of acetylcholine and glutamate overflow from the guinea pig ileum during in vitro hypoxia and hypoglycaemia. *Neurochem Int.* 2006;48(3):191–200. https://doi.org/10.1016/j.neuint.2005.10.005.
363. Swaminathan M, Hill-Yardin EL, Bornstein JC, JPP F. Endogenous glutamate excites myenteric calbindin neurons by activating group i metabotropic glutamate receptors in the mouse colon. *Front Neurosci.* 2019;13(MAY):426. https://doi.org/10.3389/fnins.2019.00426.
364. Zhang L, Wang R, Bai T, et al. EphrinB2/ephB2-mediated myenteric synaptic plasticity: mechanisms underlying the persistent muscle hypercontractility and pain in postinfectious IBS. *FASEB J.* 2019;33(12):13644–13659. https://doi.org/10.1096/fj.201901192R.
365. Liu M-T, Rothstein JD, Gershon MD, Kirchgessner AL. Glutamatergic enteric neurons. *J Neurosci.* 1997;17(12):4764–4784.
366. Tsai L. Function of GABAergic and glutamatergic neurons in the stomach. *J Biomed Sci.* 2005;12:255–266. https://doi.org/10.1007/s11373-005-1357-0.

367. Neunlist M, Michel K, Reiche D, Dobreva G, Huber K, Schemann M. Glycine activates myenteric neurones in adult guinea-pigs. *J Physiol.* 2001;536(3):727–739. https://doi.org/10.1111/j.1469-7793.2001.00727.x.
368. Chen A, Chen Y, Tang Y, et al. Hippocampal AMPARs involve the central sensitization of rats with irritable bowel syndrome. *Brain Behav.* 2017;7(3):e00650. https://doi.org/10.1002/brb3.650.
369. Smoller J. The genetics of stress-related disorders: PTSD, depression, and anxiety disorders. *Neuropsychopharmacology.* 2016;41:279–319. https://doi.org/10.1038/npp.2015.266.
370. Mumtaz F, Khan MI, Zubair M, Dehpour AR. Neurobiology and consequences of social isolation stress in animal model—a comprehensive review. *Biomed Pharmacother.* 2018;105:1205–1222. https://doi.org/10.1016/j.biopha.2018.05.086.
371. Joffe ME, Santiago CI, Engers JL, Lindsley CW, Conn PJ. Metabotropic glutamate receptor subtype 3 gates acute stress-induced dysregulation of amygdalo-cortical function. *Mol Psychiatry.* 2017;1–12. https://doi.org/10.1038/s41380-017-0015-z.
372. Chang C-H, Hsiao Y-H, Chen Y-W, Yu Y-J, Gean P-W. Social isolation-induced increase in NMDA receptors in the hippocampus exacerbates emotional dysregulation in mice. *Hippocampus.* 2015;25(4):474–485. https://doi.org/10.1002/hipo.22384.
373. Ghasemi M, Phillips C, Trillo L, De Miguel Z, Das D, Salehi A. The role of NMDA receptors in the pathophysiology and treatment of mood disorders. *Neurosci Biobehav Rev.* 2014;47:336–358. https://doi.org/10.1016/j.neubiorev.2014.08.017.
374. Jiang C, Lin W-J, Salton SR. Role of a VGF/BDNF/TrkB autoregulatory feedback loop in rapid-acting antidepressant efficacy. *J Mol Neurosci.* 2018;68: 504–509. https://doi.org/10.1007/s12031-018-1124-0.
375. Neufeld KM, Kang N, Bienenstock J, Foster JA. Reduced anxiety-like behavior and central neurochemical change in germ-free mice. *Neurogastroenterol Motil.* 2011;23 (3):255–264. https://doi.org/10.1111/j.1365-2982.2010.01620.x.
376. Savignac HM, Corona G, Mills H, et al. Prebiotic feeding elevates central brain derived neurotrophic factor, N-methyl-d-aspartate receptor subunits and d-serine. *Neurochem Int.* 2013;63(8):756–764. https://doi.org/10.1016/j.neuint.2013.10.006.
377. Gronier B, Savignac HM, Di Miceli M, et al. Increased cortical neuronal responses to NMDA and improved attentional set-shifting performance in rats following prebiotic (B-GOS®) ingestion. *Eur Neuropsychopharmacol.* 2018;28(1):211–224. https://doi.org/10.1016/j.euroneuro.2017.11.001.
378. Bäckhed F, Manchester JK, Semenkovich CF, Gordon JI. Mechanisms underlying the resistance to diet-induced obesity in germ-free mice. *Proc Natl Acad Sci U S A.* 2007;104(3):979–984. https://doi.org/10.1073/pnas.0605374104.
379. Høverstad T, Midtvedt T. Short-chain fatty acids in germfree mice and rats. *J Nutr.* 1986;116(9):1772–1776. https://doi.org/10.1093/jn/116.9.1772.
380. Pluznick JL. Microbial short-chain fatty acids and blood pressure regulation. *Curr Hypertens Rep.* 2017;19(4):25. https://doi.org/10.1007/s11906-017-0722-5.
381. Tahara Y, Yamazaki M, Sukigara H, et al. Gut microbiota-derived short chain fatty acids induce circadian clock entrainment in mouse peripheral tissue. *Sci Rep.* 2018;8 (1):1395. https://doi.org/10.1038/s41598-018-19836-7.
382. Erny D, Hrabě de Angelis AL, Prinz M. Communicating systems in the body: how microbiota and microglia cooperate. *Immunology.* 2017;150(1):7–15. https://doi.org/10.1111/imm.12645.
383. Gill PA, van Zelm MC, Muir JG, Gibson PR. Review article: short chain fatty acids as potential therapeutic agents in human gastrointestinal and inflammatory disorders. *Aliment Pharmacol Ther.* 2018;48(1):15–34. https://doi.org/10.1111/apt.14689.

384. Schönfeld P, Wojtczak L. Short- and medium-chain fatty acids in energy metabolism: the cellular perspective. *J Lipid Res*. 2016;57(6):943–954. https://doi.org/10.1194/jlr.R067629.
385. Cherbut C, Ferrier L, Rozé C, et al. Short-chain fatty acids modify colonic motility through nerves and polypeptide YY release in the rat. *Am J Physiol*. 1998;275(6): G1415–G1422. https://doi.org/10.1152/ajpgi.1998.275.6.G1415.
386. Samuel BS, Shaito A, Motoike T, et al. Effects of the gut microbiota on host adiposity are modulated by the short-chain fatty-acid binding G protein-coupled receptor, Gpr41. *Proc Natl Acad Sci U S A*. 2008;105(43):16767–16772. https://doi.org/10.1073/pnas.0808567105.
387. Dalile B, Van Oudenhove L, Vervliet B, Verbeke K. The role of short-chain fatty acids in microbiota–gut–brain communication. *Nat Rev Gastroenterol Hepatol*. 2019;16 (8):461–478. https://doi.org/10.1038/s41575-019-0157-3.
388. Arnoldussen IAC, Wiesmann M, Pelgrim CE, et al. Butyrate restores HFD-induced adaptations in brain function and metabolism in mid-adult obese mice. *Int J Obes (Lond)*. 2017;41(6):935–944. https://doi.org/10.1038/ijo.2017.52.
389. Val-Laillet D, Guérin S, Coquery N, et al. Oral sodium butyrate impacts brain metabolism and hippocampal neurogenesis, with limited effects on gut anatomy and function in pigs. *FASEB J*. 2018;32(4):2160–2171. https://doi.org/10.1096/fj.201700547RR.
390. van de Wouw M, Boehme M, Lyte JM, et al. Short-chain fatty acids: microbial metabolites that alleviate stress-induced brain-gut axis alterations. *J Physiol*. 2018;596 (20):4923–4944. https://doi.org/10.1113/JP276431.
391. Varela RB, Valvassori SS, Lopes-Borges J, et al. Sodium butyrate and mood stabilizers block ouabain-induced hyperlocomotion and increase BDNF, NGF and GDNF levels in brain of Wistar rats. *J Psychiatr Res*. 2015;61:114–121. https://doi.org/10.1016/j.jpsychires.2014.11.003.
392. Barichello T, Generoso JS, Simões LR, et al. Sodium butyrate prevents memory impairment by re-establishing BDNF and GDNF expression in experimental pneumococcal meningitis. *Mol Neurobiol*. 2015;52(1):734–740. https://doi.org/10.1007/s12035-014-8914-3.
393. Frost G, Sleeth ML, Sahuri-Arisoylu M, et al. The short-chain fatty acid acetate reduces appetite via a central homeostatic mechanism. *Nat Commun*. 2014;5:3611. https://doi.org/10.1038/ncomms4611.
394. Nagatsu T. Tyrosine hydroxylase: human isoforms, structure and regulation in physiology and pathology. *Essays Biochem*. 1995;30:15–35.
395. Zhang S, Jin Y, Zeng Z, Liu Z, Fu Z. Subchronic exposure of mice to cadmium perturbs their hepatic energy metabolism and gut microbiome. *Chem Res Toxicol*. 2015;28(10):2000–2009. https://doi.org/10.1021/acs.chemrestox.5b00237.
396. Maltz RM, Keirsey J, Kim SC, et al. Prolonged restraint stressor exposure in outbred CD-1 mice impacts microbiota, colonic inflammation, and short chain fatty acids. *PLoS One*. 2018;13(5):e0196961. https://doi.org/10.1371/journal.pone.0196961.
397. Liu S, Li E, Sun Z, et al. Altered gut microbiota and short chain fatty acids in Chinese children with autism spectrum disorder. *Sci Rep*. 2019;9(1):287. https://doi.org/10.1038/s41598-018-36430-z.
398. MacFabe DF, Cain DP, Rodriguez-Capote K, et al. Neurobiological effects of intraventricular propionic acid in rats: possible role of short chain fatty acids on the pathogenesis and characteristics of autism spectrum disorders. *Behav Brain Res*. 2007;176 (1):149–169. https://doi.org/10.1016/j.bbr.2006.07.025.
399. MacFabe DF, Cain NE, Boon F, Ossenkopp K-P, Cain DP. Effects of the enteric bacterial metabolic product propionic acid on object-directed behavior, social behavior, cognition, and neuroinflammation in adolescent rats: relevance to autism spectrum disorder. *Behav Brain Res*. 2011;217(1):47–54. https://doi.org/10.1016/j.bbr.2010.10.005.

400. Volmar C-H, Wahlestedt C. Histone deacetylases (HDACs) and brain function. *Neuroepigenetics*. 2015;1:20–27. https://doi.org/10.1016/j.nepig.2014.10.002.
401. Schroeder FA, Lin CL, Crusio WE, Akbarian S. Antidepressant-like effects of the histone deacetylase inhibitor, sodium butyrate, in the mouse. *Biol Psychiatry*. 2007;62(1):55–64. https://doi.org/10.1016/j.biopsych.2006.06.036.
402. Stafford JM, Raybuck JD, Ryabinin AE, Lattal KM. Increasing histone acetylation in the hippocampus-infralimbic network enhances fear extinction. *Biol Psychiatry*. 2012;72(1):25–33. https://doi.org/10.1016/j.biopsych.2011.12.012.
403. Brown AJ, Goldsworthy SM, Barnes AA, et al. The orphan G protein-coupled receptors GPR41 and GPR43 are activated by propionate and other short chain carboxylic acids. *J Biol Chem*. 2003;278(13):11312–11319. https://doi.org/10.1074/jbc.M211609200.
404. Nilsson NE, Kotarsky K, Owman C, Olde B. Identification of a free fatty acid receptor, FFA2R, expressed on leukocytes and activated by short-chain fatty acids. *Biochem Biophys Res Commun*. 2003;303(4):1047–1052. https://doi.org/10.1016/s0006-291x(03)00488-1.
405. Hoyles L, Snelling T, Umlai U-K, et al. Microbiome-host systems interactions: protective effects of propionate upon the blood-brain barrier. *Microbiome*. 2018;6(1):55. https://doi.org/10.1186/s40168-018-0439-y.
406. Koh A, De Vadder F, Kovatcheva-Datchary P, Bäckhed F. From dietary fiber to host physiology: short-chain fatty acids as key bacterial metabolites. *Cell*. 2016;165(6):1332–1345. https://doi.org/10.1016/j.cell.2016.05.041.
407. Braniste V, Al-Asmakh M, Kowal C, et al. The gut microbiota influences blood-brain barrier permeability in mice. *Sci Transl Med*. 2014;6(263):263ra158. https://doi.org/10.1126/scitranslmed.3009759.
408. During MJ, Cao L, Zuzga DS, et al. Glucagon-like peptide-1 receptor is involved in learning and neuroprotection. *Nat Med*. 2003;9(9):1173–1179. https://doi.org/10.1038/nm919.
409. McClean PL, Parthsarathy V, Faivre E, Hölscher C. The diabetes drug liraglutide prevents degenerative processes in a mouse model of Alzheimer's disease. *J Neurosci*. 2011;31(17):6587–6594. https://doi.org/10.1523/JNEUROSCI.0529-11.2011.
410. Painsipp E, Herzog H, Sperk G, Holzer P. Sex-dependent control of murine emotional-affective behaviour in health and colitis by peptide YY and neuropeptide Y. *Br J Pharmacol*. 2011;163(6):1302–1314. https://doi.org/10.1111/j.1476-5381.2011.01326.x.
411. Corrêa-Oliveira R, Fachi JL, Vieira A, Sato FT, Vinolo MAR. Regulation of immune cell function by short-chain fatty acids. *Clin Transl Immunol*. 2016;5(4):e73. https://doi.org/10.1038/cti.2016.17.
412. Arpaia N, Campbell C, Fan X, et al. Metabolites produced by commensal bacteria promote peripheral regulatory T-cell generation. *Nature*. 2013;504(7480):451–455. https://doi.org/10.1038/nature12726.
413. Russell DW. The enzymes, regulation, and genetics of bile acid synthesis. *Annu Rev Biochem*. 2003;72:137–174. https://doi.org/10.1146/annurev.biochem.72.121801.161712.
414. de Aguiar Vallim TQ, Tarling EJ, Edwards PA. Pleiotropic roles of bile acids in metabolism. *Cell Metab*. 2013;17(5):657–669. https://doi.org/10.1016/j.cmet.2013.03.013.
415. Maruyama T, Tanaka K, Suzuki J, et al. Targeted disruption of G protein-coupled bile acid receptor 1 (Gpbar1/M-bar) in mice. *J Endocrinol*. 2006;191(1):197–205. https://doi.org/10.1677/joe.1.06546.
416. Jiang C, Xie C, Li F, et al. Intestinal farnesoid X receptor signaling promotes nonalcoholic fatty liver disease. *J Clin Invest*. 2015;125(1):386–402. https://doi.org/10.1172/JCI76738.

417. Li F, Jiang C, Krausz KW, et al. Microbiome remodelling leads to inhibition of intestinal farnesoid X receptor signalling and decreased obesity. *Nat Commun.* 2013;4:2384. https://doi.org/10.1038/ncomms3384.
418. González-Regueiro JA, Moreno-Castañeda L, Uribe M, Chávez-Tapia NC. The role of bile acids in glucose metabolism and their relation with diabetes. *Ann Hepatol.* 2017;16(suppl 1):16–21. https://doi.org/10.5604/01.3001.0010.5672. s3–105.
419. Inagaki T, Moschetta A, Lee Y-K, et al. Regulation of antibacterial defense in the small intestine by the nuclear bile acid receptor. *Proc Natl Acad Sci U S A.* 2006;103 (10):3920–3925. https://doi.org/10.1073/pnas.0509592103.
420. Begley M, Gahan CGM, Hill C. The interaction between bacteria and bile. *FEMS Microbiol Rev.* 2005;29(4):625–651. https://doi.org/10.1016/j.femsre.2004.09.003.
421. Joyce SA, MacSharry J, Casey PG, et al. Regulation of host weight gain and lipid metabolism by bacterial bile acid modification in the gut. *Proc Natl Acad Sci U S A.* 2014;111(20):7421–7426. https://doi.org/10.1073/pnas.1323599111.
422. Vavassori P, Mencarelli A, Renga B, Distrutti E, Fiorucci S. The bile acid receptor FXR is a modulator of intestinal innate immunity. *J Immunol.* 2009;183(10): 6251–6261. https://doi.org/10.4049/jimmunol.0803978.
423. Gadaleta RM, van Erpecum KJ, Oldenburg B, et al. Farnesoid X receptor activation inhibits inflammation and preserves the intestinal barrier in inflammatory bowel disease. *Gut.* 2011;60(4):463–472. https://doi.org/10.1136/gut.2010.212159.
424. Cipriani S, Mencarelli A, Chini MG, et al. The bile acid receptor GPBAR-1 (TGR5) modulates integrity of intestinal barrier and immune response to experimental colitis. *PLoS One.* 2011;6(10):e25637. https://doi.org/10.1371/journal.pone.0025637.
425. Sarathy J, Detloff SJ, Ao M, et al. The Yin and Yang of bile acid action on tight junctions in a model colonic epithelium. *Physiol Rep.* 2017;5(10):e13294. https://doi.org/10.14814/phy2.13294.
426. Biagioli M, Carino A, Cipriani S, et al. The bile acid receptor GPBAR1 regulates the M1/M2 phenotype of intestinal macrophages and activation of GPBAR1 rescues mice from murine colitis. *J Immunol.* 2017;199(2):718–733. https://doi.org/10.4049/jimmunol.1700183.
427. Sakanaka T, Inoue T, Yorifuji N, et al. The effects of a TGR5 agonist and a dipeptidyl peptidase IV inhibitor on dextran sulfate sodium-induced colitis in mice. *J Gastroenterol Hepatol.* 2015;30(suppl 1):60–65. https://doi.org/10.1111/jgh.12740.
428. Yoneno K, Hisamatsu T, Shimamura K, et al. TGR5 signalling inhibits the production of pro-inflammatory cytokines by in vitro differentiated inflammatory and intestinal macrophages in Crohn's disease. *Immunology.* 2013;s139(1):19–29. https://doi.org/10.1111/imm.12045.
429. Liu S, Marcelin G, Blouet C, et al. A gut-brain axis regulating glucose metabolism mediated by bile acids and competitive fibroblast growth factor actions at the hypothalamus. *Mol Metab.* 2018;8:37–50. https://doi.org/10.1016/j.molmet.2017.12.003.

CHAPTER THREE

Microbiome and motor neuron diseases

Kaberi Chatterjee[a] and Sugato Banerjee[b,*]
[a]Department of Pharmaceutical Sciences and Technology, Birla Institute of Technology, Mesra, Ranchi, India
[b]Department of Pharmacology and Toxicology, National Institute of Pharmaceutical Education and Research, Kolkata, India
[*]Corresponding author: e-mail address: banerjeesugato1@gmail.com

Contents

Abstract

The microbiome is the ecological community of commensal, symbiotic, and pathogenic microorganisms that share our body space (Medical and Health Genomics, 2016, page 15–28). The human gut is the location where the maximum number of microorganisms can be found. Among the different microorganisms they can be broadly classified into two groups: the beneficial and harmful. In the human gut there is always a balance between the beneficial and the opportunistic microorganism which maintains human health. However, if the balance is not maintained and homeostasis is disturbed, with an increase in opportunistic microorganisms, it may result in various diseases like inflammatory bowel disease, irritable bowel disease, ulcerative colitis, Crohn's disease, colorectal cancer, metabolic disorders and neurodegenerative diseases including motor neuron diseases. In the present chapter we discuss the role of gut bacteria in motor neuron diseases like multiple sclerosis, Parkinson's disease and amyotrophic lateral sclerosis.

1. Introduction

The fetal gut at birth contains relatively few microorganisms, though recent research suggests that the fetal gut may not be as sterile as previously thought.[1] Slowly, as the child grows, so does the gut microbial number and its diversity. Gestational age, type of delivery (cesarean vs normal),

Progress in Molecular Biology and Translational Science, Volume 176
ISSN 1877-1173
https://doi.org/10.1016/bs.pmbts.2020.08.010

type of diet, sanitation and finally how much the child gets exposed to antibiotics determines the enteric flora.[2] The human gut is the home for both beneficial and pathogenic organisms. Among the beneficial strains Bacteriodes are the gram-negative bacteria which forms short chain fatty acids from lipopolysaccharides[3]; the family consists of: *B. fragilis*, *B. distasonis*, *B. ovatus*, *B. thetaiotaomicron*, *B. vulgatus*, *B. uniformis* and others. Bacilli are the gram-positive anaerobic bacteria. Bacillus and Clostridia form the family Bacilliaceae. Some examples of Bacillus family members include: *B. cereus*, *B. anthracis*, *B. licheniformis*, *B. subtilus*, *B. alvei*, *B. brevis*, *B. circulans*, *B. coagulans*, *B. macerans*, *B. pumilus*, *B. sphaericus*, *B. thuringiensis*. The clostridium belongs under the phylum of *Firmicutes* which is also a dominant class and present in cluster formation. The other bacteria found in gut are *Enterobacteria*, *Actenibacteria*, *Peptococci*, *Staphylococci*, and *Streptococci*, and yeast are also present in the gut. The other classes of gut bacteria and factors determining their growth have been discussed by others.[4,5]

2. Gut bacteria and Parkinson's disease

Parkinson's disease (PD) is a progressive neurodegenerative disorder where the neurodegeneration is due to misfolding of α-synuclein (α-syn) protein which leads to the formation of the Lewy bodies.[3] The accumulation of Lewy bodies in neurons leads to degenerative changes and loss of dopaminergic neurons in the basal ganglia and nigrostriatal tract resulting in the development of PD.[6]

The *Prevotellaceae*, a family in the phylum Bacteroidetes, play an important role in PD.[7–10] It is found that the amount of *Prevotellaceae* is comparatively lower in abundance than *Enterobacteriaceae* in the stool of PD patients.[11] Generally, *Prevotellaceae* produce short chain fatty acids (SCFAS) by the fermentation of dairy products.[12] Reduction in SCFAS leads to increased gut permeability increasing the intestinal exposure to bacterial endotoxins. Exposure to endotoxins leads to the misfolding of α-syn and initiates and retains its expression in the colon.[13,14] The relative abundance of Enterobacteriaceae has been correlated with gait difficulty and postural instability.[11] Lipopolysaccharide (LPS) and other neurotoxins can also alter the permeability of the intestinal wall[15] and subsequently neurotoxins are able to enter the blood stream thus increasing LPS levels in the serum of PD patients.[3] LPS is composed of three components: (1) O-antigen with multiple repeating units of monosaccharides,

(2) a polysaccharide core with a sugar (2-keto-3-deoxyoctonate), and (3) lipid which helps to form the diglucosamine backbone to which six fatty acid chains are connected. LPS is associated with the soluble LPS binding protein (LBP) and CD14,[16] together they are involved in cellular activation.[17] LPS in blood releases pro-inflammatory cytokines such as interleukin (IL)-1β, IL-6, and tumor necrosis factor (TNF)[18] and activates Toll-like receptor (TLR4) and the nuclear factor kappa B (NF-KB) pathway.[15,19] Pro-inflammatory cytokines like TNF-α, IL-1β, and IL-6 cross the blood brain barrier (BBB) and lead to α-syn deposition[20–24] in dopaminergic neurons. In the substantia nigra region, the breakdown of the BBB has been associated with degeneration of dopaminergic neurones.[25] Thus, the increase of Enterobacteriaceae and subsequent increase the amount of LPS is responsible for PD progression.[26,27] Blautia, Coprococcus, Roseburia and Faecalibacterium, which are the butyrate producing bacteria with anti-inflammatory properties, are found in PD patients.[28] Excessive bacterial growth (SIBO) in the small intestine which leads to dysbiosis is a cause of motor impairment in PD.[29,30] Helicobacter pylori infection can also act as a triggering factor in PD.[31]

In multiple system atrophy (MSA) and Steele-Richardson-Olszewski syndrome, the changes in the gut bacteria composition is similar to atypical PD[32] with a significant (43%) decrease of *Lachnospiraceae* compared to Bifidobacteria.[33]

Gut bacteria have also been shown to modulate oxidative stress induced changes in the PD brain. Nicotinamide adenine dinucleotide (NAD) can generate energy to maintain the integrity of DNA, ATP, calcium channel signaling along with apoptosis, and can generate reactive oxygen species (ROS).[17] The rat model of 6-hydroxy dopamine induced PD, with neurodegenerative changes in the ventral tegmental area (VTA) and SN, has been associated with depletion of superoxide dismutase and glutathione peroxidase thus linking oxidative stress and PD.[34] Malondialdehyde, a lipid peroxidation product that has been reported to increase in LPS treated rats, when combined with 6-hydroxy dopamine showed a more pronounced manifestation of PD.[35] Besides ROS, reactive nitrogen species (RNS), entities produced by the interactions between superoxide and nitric oxide (NO), play an important role in the pathogenesis of PD. Studies using post-mortem PD brains showed iNOS and nNOS expression, highlighting the role of NO in this neurodegenerative disorder.[36–39] Oxidative stress induced by menadione and H_2O_2 deregulates the redox ratio by decreasing folate synthesis of *Enterococcus durans* in colorectal cancer.[40] While another study showed the association between hydrogen production by both gram-negative and gram-positive gut bacteria in the progression of PD.[41]

Gut bacteria like enterococci which harbor tyrosine decarboxylases, have been shown to decarboxylate levodopa to dopamine in the small intestine. PD patients could have high abundance of these bacteria due to small intestinal overgrowth (SIBO). When SIBO is eradicated in PD patients using rifaximin, motor fluctuations were improved without changing the pharmacokinetics of L-Dopa.[29] Helicobacter pylori has also been reported to affect the levodopa absorption, one possible explanation is by alteration of gastric acidity, thus altering the pharmacokinetics of levodopa, delavirdine, and thyroxine.[42] One in vitro study showed that adhesins on the outer membrane of Helicobacter pylori can bind to levodopa thus altering its bioavailability in Helicobacter pylori infected PD patients.

Thus, from the given literature we may conclude that a leaky gut promotes more LPS entry into the blood, leading to the release of pro-inflammatory cytokines which may activate TLR4 and NF-KB pathways. NF-KB in turn may activate certain genes which are involved in the activation of more pro-inflammatory cytokines like TNF-α, IL-1β, and IL-6 which cross the BBB and facilitate α-syn aggregation. Butyrate producing bacteria may prevent Parkinson's while excess pathogenic bacteria in the gut have been shown to facilitate the disease. Gut bacteria may also modulate oxidative and nitrosative stress associated neuronal damage in Parkinson's disease. While intestinal microbes may also affect the absorption and metabolism of L-Dopa thus affecting Parkinson's therapy.

3. Gut bacteria and multiple sclerosis

Multiple sclerosis (MS) is autoimmune chronic neurological disorder of the central nervous system (CNS) due to generating autoantibodies and auto-reactive T cells.[43] This leads to focal inflammation in the myelin sheath and axonal degeneration resulting in dysfunctional neurons and oligodendrocytes.[44,45] Currently 2.5 million MS patients are diagnosed worldwide.[46] The average life span of MS patient is 5–10 years less than the normal population.[47] Females are more prone to the MS than males, but the manifestation is more severe and recovery is less in males than females.[48] Symptoms of MS include gait instability, sensory disturbances, fatigue, weakness, and spasticity, loss of balance, tremor, visual problems, depression, urinary bladder and bowel dysfunction.[49–51] MS can be primarily divided into Progressive Relapsing MS (PRMS), Primary progressive MS (PPMS),

Secondary progressive MS (SPMS) and Relapsing remitting MS (RRMS). Among the above RRMS is the most common form of MS.[52] In the case of RRMS there is a decrease in the anti-inflammatory responses through regulatory T cells which in turn may be modulated by IL-10.[49–51]

Various reports have shown altered gut microbial population in individuals with MS.[53] The number of colonies of *Parabacteriodes distasonis* and *Prevotella copi*[54] is decreased in MS; these bacteria produce anti-inflammatory responses whereas the colonies of *Firmicutes* spp., Clostidium and Faccalibacterium are increased in MS.[55] In MS patients there is an increase of *Acinetobacter calcoaceticus* which stimulates pro-inflammatory cytokines, Th1 and blocks the regulatory T cells thus facilitating the disease.[56] *Clostedium perfringes* type B may increase the amount of epsilon toxin which degrades the blood brain barrier (BBB) and damages neurons and oligodendrocytes leading to autoimmune demyelinating events.[57–59] Some species of Dorea which can stimulate TNF-α, metabolize sialic acid and degrade mucin, and have been reported in MS patients.[60] While *Akkermansia muciniphila* and *Blautia coccoides*, which are increased in the MS gut,[61,62] degrade mucin to short chain fatty acids.[63] They both have been implicated in both immunoregulatory and pro-inflammatory pathways, the latter being involved in upregulation of genes associated with antigen-presentation, B and T cell activation and signaling and complement activation.[64] Human studies also show higher abundance of 16S ribosomal RNA in MS patients and a significant decrease in butyrate production and T cell regulation. It is also found that probiotic treatment can help in MS patients as it has anti-inflammatory actions and reduces the peripheral immune response.[65–67]

Gut dysbiosis, or the imbalance of gut bacteria, can modulate MS pathophysiology by various mechanisms. Gut dysbiosis increases intestinal permeability which facilitates the entry of LPS. LPS ultimately activates ceramide, which can alter mitochondrial function and increases apoptotic susceptibility. The attenuation of gut butyrate production contributes to suboptimal mitochondrial function, which increases apoptotic susceptibility, as well as the reactivity of immune cells, glia, and platelets, contributing to a wider pro-inflammatory milieu. Butyrate plays an important role in maintaining the integrity of the gut barrier thus preventing the leaky gut associated increase in circulating LPS which may produce oxidative stress and pro-inflammatory activity, in turn affecting mitochondria function.[68] When oligodendrocyte and microglia cells were co-cultured as an in vitro model of MS, LPS was found to decrease oligodendrocyte myelin production via microglia activation. LPS increased microglial inducible nitric oxide synthase (iNOS) and NADPH oxidase production resulting in peroxynitrite mediated oxidative and nitrosative stress which was believed to affect oligodendrocyte mitochondrial function.[69] Butyrate has been shown to reduce the actions of LPS in various cell types,[70–73] including microglia.[74] Circulating butyrate can modulate immune cells and dampen glia-activation thus reducing inflammation.[75] It has been suggested that loss of gut-derived circulating butyrate in MS may be responsible for the effect of LPS on microglia and suppression of oligodendrocyte mitochondrial function. Butyrate has been shown to inhibit histone deacetylase (HDAC) which may help in the differentiation of immature oligodendrocytes and also optimize mitochondrial function.[68,71,73] Thus, butyrate producing gut bacteria may affect myelin synthesis by normalizing mitochondrial function and reducing microglia activation.

Gut bacteria can modulate the pathogenesis of MS by various pathways which can be summarized as follows. *Acinetobacter calcoaceticus* which stimulates pro-inflammatory cytokines, Th1 and blocks the regulatory T cells have been shown to facilitate MS. Type B *Clostedium perfringes* may increase the amount of epsilon toxin which degrades the BBB in MS. While *Blautia coccoides* have been implicated in MS by both immunoregulatory and pro-inflammatory pathways, including upregulation of genes associated with antigen-presentation, B and T cell activation as well as complement activation. Like other diseases, in MS, gut dysbiosis increases intestinal permeability which facilitates the entry of LPS. Besides promoting peripheral inflammation, LPS promotes oxidative and nitrosative stress which alone or through activation of ceramide can mediate altered mitochondrial function and increase apoptotic susceptibility to oligodendrocytes. Butyrate plays an important role in maintaining the integrity of the gut barrier. Circulating butyrate can also modulate immune cells and dampen glial activation thus reducing inflammation mediated damage in MS. Butyrate can also inhibit HDAC which may help in the differentiation of immature oligodendrocytes and optimize the mitochondrial function, helping in the normal functioning of oligodendrocytes.

4. Gut bacteria and amyotrophic lateral sclerosis

Amyotrophic lateral sclerosis (ALS) is a progressive neurodegenerative disorder which leads to the death of motor neurons in the brain and spinal cord.[3] Early stages of the disease are characterized by muscle weakness, stiffness, spasms, and cramping.[3] With time, these symptoms lead to difficulties in speaking, swallowing and eventually leads to breathing difficulties.[3] ALS may be familial and sporadic, with sporadic being more common but of an unknowing.[76] In ALS animal models, it is found that during the pre-onset stage there is an increase in gut permeability and a breakdown of the blood brain barrier (BBB).[3] The damage to the wall of the intestine is due to the decreased amount of zonula occludens (ZO)-1 and *E*-Cadherin[3,77] which are important for maintaining the integrity of the intestinal wall.[77]

The membrane integrity of BBB and BSCB (blood spinal cord barrier) is reduced in ALS patients [77a,77b] along with a reduction in the expression of ZO-1, JAM-1 (junction adhesion molecule), VE-cadherin (vascular endothelial) and claudi-5 (Garbuzova-Davis et al., 2012). These changes cause an increase in the permeability of the BBB. Leakiness in the BBB causes infiltration of immune cells into the CNS (Bataveljic et al., 2014). ALS the brain and spinal cord, and also leads to increased infiltration of macrophages, mast cells and COX 2,[77c]which may all contribute to neuroinflammation associated neurodegeneration in the ALS brain.

In the feces of SOD1G93A ALS mice, the amount of Butyrivibrio fibrosolvens and Peptostreptococcus is reduced, resulting in an increase in serum IL-7 which correlates with the decrease in butyrate levels.[78,79] Increases in butyrate reduces the progression of the ALS.[80] Fecal samples of ALS patients contain high concentrations of different species of Dorea which leads to the synthesis of ethanol from the metabolism of glucose[79]; there is also a decrease in different butyrate producing bacteria like Oscillibactor, Anacrostipes and Lachnospira species.[81] The alteration of LPS and β-*N*-methylamino-L-alanine (BMAA) in plasma which lead to neuroinflammation and breakdown of blood brain barrier (BBB) function[82–84] has been reported in the plasma of ALS individuals.

Thus, based on the current literature on gut microbiota and ALS we may conclude that gut dysbiosis may alter the integrity of the BBB by reducing the expression of tight junction proteins allowing the entry to LPS and immune cells which mediates neuroinflammation and eventual neurodegeneration.

5. Conclusion

Gut dysbiosis leads to increases in non-beneficial and opportunistic microorganisms. These organisms increase the levels of LPS and BMAA which in turn leads to disruption of gut permeability (due to damage of tight

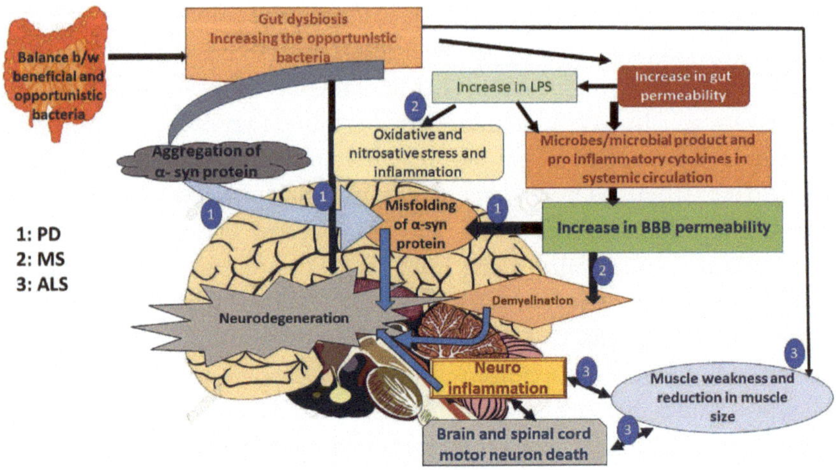

Fig. 1 Role of gut dysbiosis in the neuropathogenesis of PD, ALS and MS. Gut dysbiosis leads to increases in non-beneficial and opportunistic microorganisms. These organisms increase the levels of LPS and BMAA which in turn leads to disruption of gut permeability (due to damage of ZO-1 tight junction proteins) and increases in systemic circulation of microbes, microbial products, and pro-inflammatory factors which in turn disrupt BBB (mediated by TH1, TH17, monocytes and B cells) and contribute to neuroinflammation and neurodegeneration. α-synuclein aggregation in enteric nervous system is enhanced due to gut dysbiosis. They all contribute toward demyelination, axonal damage and neuronal death. LPS, lipopolysaccharide; BMAA, β-*N*-methylamino-L-alanine; α-Syn, α-synuclein; BBB, blood brain barrier; ENS, enteric nervous system; ALS, amyloid lateral sclerosis; MS, multiple sclerosis; PD, Parkinson's disease.

junction proteins) and increases in systemic circulation of microbial products like LPS. They in turn induce pro-inflammatory factors (cytokines like TNF-α, IL-17, IL-6, IL-1β) which may be involved in TH1, monocytes and B cell mediated disruption of the BBB and contribute to neuroinflammation and neurodegeneration (Fig. 1). The above mediators alone or in combination may contribute toward α-synuclein aggregation, demyelination, axonal damage and neuronal death leading to motor neuron disorders.

References

1. Stinson LF, Boyce MC, Payne MS, Keelan JA. The not-so-sterile womb: evidence that the human fetus is exposed to bacteria prior to birth. *Front Microbiol.* 2019;10:1124.
2. Quigley EM. Gut bacteria in health and disease. *Clin Gastroenterol Hepatol.* 2013; 9(9):560.
3. Sarkar SR, Banerjee S. Gut microbiota in neurodegenerative disorders. *J Neuroimmunol.* 2019;328:98–104.
4. Bruce-Keller A, Salbaum JM, Berthoud HR. Harnessing gut microbes for mental health: getting from here to there. *Biol Psychiatry.* 2018;83(3):214–223.

5. Valdes AM, Walter J, Segal E, Spector TD. Role of the gut microbiota in nutrition and health. *BMJ*. 2018;361:k2179.
6. Dickson DW, Fujishiro H, Orr C, et al. Neuropathology of non-motor features of Parkinson disease. *Parkinsonism Relat Disord*. 2009;15:S1–S5.
7. Braak H, de Vos RA, Bohl J, Del Tredici K. Gastric α-synuclein immunoreactive inclusions in Meissner's and Auerbach's plexuses in cases staged for Parkinson's disease-related brain pathology. *Neurosci Lett*. 2006;396(1):67–72.
8. Kieburtz K, Wunderle KB. Parkinson's disease: evidence for environmental risk factors. *Mov Disord*. 2013;28(1):8–13.
9. Shannon KM, Keshavarzian A, Dodiya HB, Jakate S, Kordower JH. Is alphasynuclein in the colon a biomarker for premotor Parkinson's disease? Evidence from 3 cases. *Mov Disord*. 2012;27(6):716–719.
10. Savica R, Carlin JM, Grossardt BR, et al. Medical records documentation of constipation preceding Parkinson disease a case-control study. *Neurology*. 2009;73(21):1752–1758.
11. Scheperjans F, Aho V, Pereira PA, et al. Gut microbiota are related to Parkinson's disease and clinical phenotype. *Mov Disord*. 2015;30(3):350–358.
12. Arumugam M, Raes J, Pelletier E, et al. Enterotypes of the human gut microbiome. *Nature*. 2011;473(7346):174.
13. Forsyth CB, Shannon KM, Kordower JH, et al. Increased intestinal permeability correlates with sigmoid mucosa alphasynuclein staining and endotoxin exposure markers in early Parkinson's disease. *PLoS One*. 2011;6(12):e28032.
14. Niehaus I, Lange JH. Endotoxin: is it an environmental factor in the cause of Parkinson's disease? *Occup Environ Med*. 2003;60(5):378.
15. Guo S, Al-Sadi R, Said HM, Ma TY. Lipopolysaccharide causes an increase in intestinal tight junction permeability in vitro and in vivo by inducing enterocyte membrane expression and localization of TLR-4 and CD14. *Am J Pathol*. 2013;182(2):375–387.
16. Dutta G, Zhang P, Liu B. The lipopolysaccharide Parkinson's disease animal model: mechanistic studies and drug discovery. *Fundam Clin Pharmacol*. 2008;22(5):453–464.
17. Ilie OD, Ciobica A, McKenna J, Doroftei B, Mavroudis I. Minireview on the relations between gut microflora and Parkinson's disease: further biochemical (oxidative stress), inflammatory, and neurological particularities. *Oxid Med Cell Longev*. 2020;2020: 4518023.
18. Zhang JM, An J. Cytokines, inflammation and pain. *Int Anesthesiol Clin*. 2007;45(2):27.
19. Tufekci KU, Genc S, Genc K. The endotoxin-induced neuroinflammation model of Parkinson's disease. *Parkinsons Dis*. 2011;2011, 487450.
20. Chen H, O'Reilly EJ, Schwarzschild MA, Ascherio A. Peripheral inflammatory biomarkers and risk of Parkinson's disease. *Am J Epidemiol*. 2007;167(1):90–95.
21. Block ML, Zecca L, Hong JS. Microglia-mediated neurotoxicity uncovering the molecular mechanisms. *Nat Rev Neurosci*. 2007;8(1):57–69.
22. Reale M, Iarlori C, Thomas A, et al. Peripheral cytokines profile in Parkinson's disease. *Brain Behav Immun*. 2009;23(1):55–63.
23. Dobbs RJ, Charlett A, Purkiss AG, Dobbs SM, Weller C, Peterson DW. Association of circulating TNF-alpha and IL-6 with ageing and parkinsonism. *Acta Neurol Scand*. 1999;100(1):34–41.
24. Sui YT, Bullock KM, Eickson MA, Zhang J, Banks WA. Alpha synuclein is transported into and out of the brain by the blood-brain barrier. *Peptides*. 2014;62:197–202.
25. Rite I, Machado A, Cano J, Venero JL. Blood-brain barrier disruption induces in vivo degeneration of nigral dopaminergic neurons. *J Neurochem*. 2007;101(6):1567–1582.
26. Mulak A, Bonaz B. Brain-gut-microbiota axis in Parkinson's disease. *World J Gastroenterol*. 2015;21(37):10609–10620.
27. Nair AT, Ramachandran V, Joghee NM, Antony S, Ramalingam G. Gut microbiota dysfunction as reliable non-invasive early diagnostic biomarkers in the pathophysiology of Parkinson's disease: a critical review. *J Neurogastroenterol Motil*. 2018;24(1):30–42.

28. Keshavazian A, Green SJ, Engen PA, et al. Colonic bacterial composition in Parkinson's disease. *Mov Disord.* 2015;30(10):1351–1360.
29. Fasano A, Bove F, Gabrielli M, et al. The role of small intestinal bacterial overgrowth in Parkinson's disease. *Mov Disord.* 2013;28(9):1241–1249.
30. Tan AH, Mahadeva S, Thalha AM, et al. Small intestinal bacterial overgrowth in Parkinson's disease. *Parkinsonism Relat Disord.* 2014;20(5):535–540.
31. Çamcı G, Oğuz S. Association between Parkinson's disease and Helicobacter Pylori. *J Clin Neurol.* 2016;12(2):147–150.
32. Barichella M, Severgnini M, Cilia R, et al. Unraveling gut microbiota in Parkinson's disease and atypical parkinsonism. *Mov Disord.* 2019;34(3):396–405.
33. Lin A, Zheng W, He Y, et al. Gut microbiota in patients with Parkinson's disease in southern China. *Parkinsonism Relat Disord.* 2018;53:82–88.
34. Hritcu L, Ciobica A, Artenie V. Effects of right-unilateral 6-hydroxydopamine infusion-induced memory impairment and oxidative stress: relevance for Parkinson's disease. *Open Life Sci.* 2008;3(3):250–257.
35. Hritcu L, Ciobica A. Intranigral lipopolysaccharide administration induced behavioral deficits and oxidative stress damage in laboratory rats: relevance for Parkinson's disease. *Behav Brain Res.* 2013;253:25–31.
36. Carr AC, McCall MR, Frei B. Oxidation of LDL by myeloperoxidase and reactive nitrogen species: reaction pathways and antioxidant protection. *Arterioscler Thromb Vasc Biol.* 2000;20(7):1716–1723.
37. Szabó C, Ischiropoulos H, Radi R. Peroxynitrite: biochemistry, pathophysiology and development of therapeutics. *Natl Rev.* 2007;6(8):662–680.
38. Hirsch EC, Breidert T, Rousselet E, Hunot S, Hartmann A, Michel PP. The role of glial reaction and inflammation in Parkinson's disease. *Ann N Y Acad Sci.* 2003;991:214–228. documentation page.
39. Malkus KA, Tsika E, Ischiropoulos H. Oxidative modifications, mitochondrial dysfunction, and impaired protein degradation in Parkinson's disease: how neurons are lost in the Bermuda triangle. *Mol Neurodegener.* 2009;4(1):24.
40. Jose S, Bhalla P, Suraishkumar GK. Oxidative stress decreases the redox ratio and folate content in the gut microbe, enterococcus durans (MTCC 3031). *Sci Rep.* 2018;8(1):1–7.
41. Suzuki A, Ito M, Hamaguchi T, et al. Quantification of hydrogen production by intestinal bacteria that are specifically dysregulated in Parkinson's disease. *PLoS One.* 2018; 13(12):e0208313.
42. Lahner E, Annibale B, Fave GD. Systematic review: Helicobacter pylori infection and impaired drug absorption. *Aliment Pharmacol Ther.* 2009;29(4):379–386.
43. Wootla B, Eriguchi M, Rodriguez M. Is multiple sclerosis an autoimmune disease? *Autoimmune Dis.* 2012;16:2012.
44. Fernando V, Omura S, Sato F, et al. Regulation of an autoimmune model for multiple sclerosis in Th2-biased GATA3 transgenic mice. *Int J Mol Sci.* 2014;15:1700–1718.
45. Zhang X, Tao Y, Chopra M, et al. IL-11 induces Th17 cell responses in patients with early relapsing-remitting multiple sclerosis. *J Immunol Res.* 2015;194:5139.
46. Evans E, Piccio L, Cross AH. Use of vitamins and dietary supplements by patients with multiple sclerosis: a review. *JAMA Neurol.* 2018;75:1013–1021.
47. Lin R, Charlesworth J, van der Mei I, Taylor BV. The genetics of multiple sclerosis. *Pract Neurol.* 2012;12:279–288.
48. Ribbons KA, McElduff P, Boz C, et al. Male sex is independently associated with faster disability accumulation in relapse-onset ms but not in primary progressive mS. *PLoS One.* 2015;10:e0122686.
49. Chu F, Shi M, Lang Y, et al. Gut microbiota in multiple sclerosis and experimental autoimmune encephalomyelitis: current applications and future perspectives. *Mediators Inflamm.* 2018;2018:8168717.

50. Loma I, Heyman R. Multiple sclerosis: pathogenesis and treatment. *Curr Neuropharmacol.* 2011;9:409–416.
51. van den Hoogen WJ, Laman JD, Hart BA. Modulation of multiple sclerosis and its animal model experimental autoimmune encephalomyelitis by food and gut microbiota. *Front Immunol.* 2017;8:1081.
52. Miljković D, Spasojević I. Multiple sclerosis: molecular mechanisms and therapeutic opportunities. *Antioxid Redox Signal.* 2013;19(18):2286–2334.
53. Kirby TO, Ochoa-Repáraz J. The gut microbiome in multiple sclerosis: a potential therapeutic avenue. *Med Sci.* 2018;6(3):69.
54. Miyake S, Kim S, Suda W, et al. Dysbiosis in the gut microbiota of patients with multiple sclerosis, with a striking depletion of species belonging to clostridia XIVa and IV clusters. *PLoS One.* 2015;10:e0137429.
55. Chen J, Chia N, Kalari KR, et al. Multiple sclerosis patients have a distinct gut microbiota compared to healthy controls. *Sci Rep.* 2016;6:28484.
56. Cekanaviciute E, Yoo BB, Runia TF, et al. Gut bacteria from multiple sclerosis patients modulate human T cells and exacerbate symptoms in mouse models. *Proc Natl Acad Sci U S A.* 2017;114(40):10713–10718.
57. Lonchamp E, Dupont JL, Wioland L, et al. *Clostridium perfringens* epsilon toxin targets granule cells in the mouse cerebellum and stimulates glutamate release. *PLoS One.* 2010;5(9):e13046.
58. Dorca-Arévalo J, Soler-Jover A, Gibert M, Popoff MR, Martín-Satué M, Blasi J. Binding of epsilon-toxin from Clostridium perfringens in the nervous system. *Vet Microbiol.* 2008;131(1–2):14–25.
59. Finnie JW, Blumbergs PC, Manavis J. Neuronal damage produced in rat brains by clostridium perfringens type D epsilon toxin. *J Comp Pathol.* 1999;120:415–420.
60. Schirmer M, Smeekens SP, Vlamakis H, et al. Linking the human gut microbiome to inflammatory cytokine production capacity. *Cell.* 2016;167(4):1125–1136.
61. Cekanaviciute E, Debelius JW, Singh S, et al. Gut dysbiosis isa feature of MS and it is characterized by bacteria able to regulate lymphocyte differentiation in vitro. *Mult Scler J.* 2016;22(S3):58–59.
62. Jangi S, Gandhi R, Cox LM, et al. Alterations of the human gut microbiome in multiple sclerosis. *Nat Commun.* 2016;7:12015.
63. Hughes LE, Smith PA, Bonell S, et al. Cross-reactivity between related sequences found in Acinetobacter sp., *Pseudomonas aeruginosa*, myelin basic protein and myelin oligodendrocyte glycoprotein in multiple sclerosis. *J Neuroimmunol.* 2003;144(1–2):105–115.
64. Derrien M, Van Baarlen P, Hooiveld G, Norin E, Muller M, de Vos W. Modulation of mucosal immune response, tolerance, and proliferation in mice colonized by the mucin-degrader Akkermansia muciniphila. *Front Microbiol.* 2011;2:166.
65. Eisen A, Greenberg BM, Bowen JD, Arnold DL, Caggiano AO. A double-blind, placebo-controlled, single ascending-dose study of remyelinating antibody rHIgM22 in people with multiple sclerosis. *Mult Scler J Exp Transl Clin.* 2017;3(4):2055217317743097.
66. Chami M, Halmer R, Schnoeder L, et al. Acid sphingomyelinase deficiency enhances myelin repair after acute and chronic demyelination. *PLoS One.* 2017;12(6):e0178622.
67. Podbielska M, Szulc ZM, Kurowska E, et al. Cytokine-induced release of ceramide-enriched exosomes as a mediator of cell death signaling in an oligodendroglioma cell line. *J Lipid Res.* 2016;57(11):2028–2039.
68. Anderson G. Gut dysbiosis dysregulates central and systemic homeostasis via decreased melatonin and suboptimal mitochondria functioning: pathoetiological and pathophysiological implications. *Melatonin Res.* 2019;2:70–85.
69. Fetisova EK, Muntyan MS, Lyamzaev KG, Chernyak BV. Therapeutic effect of the mitochondria-targeted antioxidant skq1 on the culture model of multiple sclerosis. *Oxid Med Cell Longev.* 2019;2019, 2082561.

70. Roy AC, Chang G, Ma N, et al. Sodium butyrate suppresses NOD1-mediated inflammatory molecules expressed in bovine hepatocytes during iE-DAP and LPS treatment. *J Cell Physiol.* 2019;234(11):19602–19620.
71. Chen J, Wu Y, Sun Y, et al. Bacterial endotoxin decreased histone H3 acetylation of bovine mammary epithelial cells and the adverse effect was suppressed by sodium butyrate. *BMC Vet Res.* 2019;15(1):267.
72. Liu J, Chang G, Huang J, et al. Sodium butyrate inhibits the inflammation of lipopolysaccharide-induced acute lung injury in mice by regulating the toll-like receptor 4/nuclear factor κB signaling pathway. *J Agric Food Chem.* 2019;67(6):1674–1682.
73. Chen T, Noto D, Hoshino Y, Mizuno M, Miyake S. Butyrate suppresses demyelination and enhances remyelination. *J Neuroinflammation.* 2019;16(1):165.
74. Yamawaki Y, Yoshioka N, Nozaki K, et al. Sodium butyrate abolishes lipopolysaccharide-induced depression-like behaviors and hippocampal microglial activation in mice. *Brain Res.* 2018;1680:13–38.
75. Park J, Wang Q, Wu Q, Mao-Draayer Y, Kim CH. Bidirectional regulatory potentials of short-chain fatty acids and their G-protein-coupled receptors in autoimmune neuroinflammation. *Sci Rep.* 2019;9(1):1–3.
76. Toepfer M, Schroeder M, Klauser A, et al. Delayed colonic transit times in amyotrophic lateral sclerosis assessed with radio-opaque markers. *Eur J Med Res.* 1997;2(11):473–476.
77. Zhou Y, Lu Y, Fang X, et al. An astrocyte regenerative response from vimentin-containing cells in the spinal cord of amyotrophic lateral sclerosis's disease-like transgenic (G93A SOD1) mice. *J Neurodegener Dis.* 2015;15(1):1–2. (a) Bataveljic D, Milosevic M, Radenovic L, Andjus P. Novel molecular biomarkers at the blood-brain barrier in ALS. *BioMed Res Int.* 2014;2014:907545. (b) Garbuzova-Davis S, Hernandez-Ontiveros DG, Rodrigues MCO, et al. Impaired blood-brain/spinal cord barrier in ALS patients. *Brain Res.* 2012;1469:114–128. (c) Graves M, Fiala M, Dinglasan LA, et al. Inflammation in amyotrophic lateral sclerosis spinal cord and brain is mediated by activated macrophages, mast cells and T cells. *Amyotroph Lateral Scler Other Motor Neuron Disord.* 2004; 5(4):213–219.
78. Wu S, Yi J, Zhang YG, Zhou J, Sun J. Leaky intestine and impaired microbiome in an amyotrophic lateral sclerosis mouse model. *Physiol Rep.* 2015;3(4):12356.
79. Fang X. Potential role of gut microbiota and tissue barriers in Parkinson's disease and amyotrophic lateral sclerosis. *Int J Neurosci.* 2015;126(9):771–776.
80. Zhang YG, Wu S, Yi J, et al. Target intestinal microbiota to alleviate disease progression in amyotrophic lateral sclerosis. *Clin Ther.* 2017;39(2):322–336.
81. Fang X, Wang X, Yang S, et al. Evaluation of the microbial diversity in amyotrophic lateral sclerosis using high-throughput sequencing. *Front Microbiol.* 2016;7:1479.
82. Wright ML, Fournier C, Houser MC, Tansey M, Glass J, Hertzberg VS. Potential role of the gut microbiome in ALS: a systematic review. *Biol Res Nurs.* 2018;20(5):1–9.
83. Catanzaro R, Anzalone M, Calabrese F, et al. The gut microbiota and its correlations with the central nervous system disorders. *Panminerva Med.* 2015;57(3):127–143.
84. Bengmark S. Gut microbiota, immune development and function. *Pharmacol Res.* 2013;69(1):87–113.

CHAPTER FOUR

The gut microbiome and psycho-cognitive traits

Nicholas A. Vernice[a], Neal Shah[b], Eric Lam[c], Pamela Herd[d], Allison B. Reiss[b,e], and Lora J. Kasselman[e,*]

[a]Weill Cornell Medicine, New York, NY, United States
[b]NYU Winthrop Hospital, Mineola, NY, United States
[c]Nassau University Medical Center, East Meadow, NY, United States
[d]McCourt School of Public Policy, Georgetown University, Washington, DC, United States
[e]NYU Long Island School of Medicine, Mineola, NY, United States
[*]Corresponding author: e-mail address: lora.kasselman@nyulangone.org

Contents

Abstract

The idea that trillions of bacteria inhabit our gut is somewhat unnerving, yet these bacteria may have a greater influence on our behavior than previously thought. Accumulating data strongly suggest that these gut commensal organisms have a strong inter-relationship with our brain and behavior, including cognitive function, mood, and personality. In this chapter, we discuss the role of the gut microbiome in the development of human personality, mood and mood disorders, and cognition, with a particular emphasis on the current consensus and controversies in the literature surrounding the behavioral effects of bioactive metabolites, microbial ratio shifts, and neurotransmitter synthesis facilitated by the microbiome.

Progress in Molecular Biology and Translational Science, Volume 176
ISSN 1877-1173
https://doi.org/10.1016/bs.pmbts.2020.08.014

1. Introduction

Given the longstanding associations between human thoughts and feelings and gastrointestinal discomfort, it is not surprising that research over the past decade has succeeded in elucidating a continually evolving link between gut microbial composition and psychocognitive traits such as personality, mood, and cognition. This connection leads to questions regarding which influences which, i.e. does the gut composition influence psychocognitive traits or does our brain shape our behavior, including feeding behavior, which then determines gut compositionality? Most likely, the answer lies in the middle with bi-directional communication between the gut microbiome and the brain leading to commensurate changes in both places, potentially starting before birth. Despite this somewhat discomforting idea that microbes in our gut may be influencing our brain, the importance of this interaction needs to be highlighted and further studied due to the beneficial effects seen in recent studies. In particular, the gut microbiome's ability to attenuate and initiate immune responses, synthesize neurotransmitters, alter gut permeability, and produce bioactive metabolites suggests an overwhelmingly complex and fascinating new inter-kingdom system that may assist in the formation of the very essence of our being by shaping of our personality, affecting our mood, and influencing our cognitive abilities. This chapter will focus on research involving the gut microbiome and personality, mood, and cognition.

2. Microbiome and personality

The term "personality" originates from the Latin word *persona*, which is a theatrical mask worn by an actor on the stage to represent and project specific traits of a character. As such, personality refers to the enduring characteristic sets of cognitive, emotional, and behavioral patterns that distinguish one person from another within a society.[1] Studies in personality psychology provide classification systems of personalities utilizing self-report data to distinguish personality types and temperaments,[2–4] however these studies allow only limited explanations and insights into why individuals differ. Abundant evidence from recent studies suggest that both nature (genes) and nurture (environment) contribute to the development of personality.[5] The microbiome is identified as one of the environmental factors that help shape the personality of an individual.

Both animal and human studies have shown associations between gut microbes and different aspects of personality. While animal models show consistent relationships between the gut microbiome and anxiety and fear, personality, per se, is harder to measure in a way that's analogous to human personality when employing animal models.[6] A few animal studies, however, provide intriguing hypotheses as to the relationships between the gut microbiome and temperament. For example, in a study of maternal-diet-induced gut dysbiosis, offspring with reduced gut diversity showed impaired sociability with other mice.[7] Similarly, mice pups with altered gut composition in response to maternal diet show reduced exploratory behavior later in life.[8]

Human studies on the relationships between the gut microbiome and personality are also not extensive, but provide exciting evidence. The largest adult study, on a non randomly selected sample of South Koreans, found relationships between the gut microbiome and personality.[9] In this sample of just over 600 South Koreans, there was evidence of relationships between the gut microbiome and, especially, neuroticism and conscientiousness. The sample, however, was a convenience sample that disproportionately included men. Given sex differences in both personality and the gut microbiome, this may have affected the findings. Moreover, as the authors note, it's not clear whether this would replicate in populations outside of South Korea, such as the American context where cultural differences play an important role in personality differences.

The remaining human studies have focused on infants and toddlers, which provide compelling findings, but given variance in how early life temperament predicts subsequent life course personality, the comparability to adults is limited. In a non-random sample of 60 healthy human adults, different gut microbial enterotypes were associated with temperament, but these data did not show sex differences in gut composition which is unusual. Nonetheless, results from a study on 79 mother-toddler pairs showed that toddlers with greater phylogenetic diversity of gut microbiota displayed more mother-rated temperament traits such as extraversion.[10] Further, In a sample of 301 young infants drawn from a Finnish cohort, gut alpha diversity showed a negative association with fear reactivity and negative emotion while positive associations were found between Bifidobacterium and Streptococcus genera and positive emotionality, as measured by maternal report.[11]

Due to its complexity and lack of objective quantification, personality may be one of the most difficult psycho-cognitive traits to study with respect to the microbiome. The recent advancement in personality neuroscience utilizing neuroimaging such as functional magnetic resonance imaging (MRI) and positron emission tomography, electroencephalogram, molecular genetics, and biochemical analysis of neurotransmitters offers novel understanding of the relationship between brain connectivity and personality. In one recent functional MRI study, alpha diversity of the gut microbiome was associated with connectivity between brain areas important in processing and responding to threat.[12] Future studies should continue to incorporate these technologies to investigate the impact of microbiome in personality development.

3. Microbiome and mood

3.1 Overview

The fifth edition of the Diagnostic and Statistical Manual of Mental Disorders (DSM-V) defines "mood" as a "pervasive and sustained emotional 'climate,' differentiating a patient's mood from their more immediate and transient *affect*," or "pattern of observable behaviors that is the expression of a subjectively experienced feeling state."[13] The Mental Status Examination (MSE), now a hallmark of the psychiatric interview, places a strong emphasis on analysis of the patient's mood and affect, with observations regarding these terms being integral to the assessment and diagnosis of behavioral and mood disorders.[14] Interestingly, diagnoses of gastrointestinal pathologies have long been comorbid with those of mood disorders.[15] Within the last decade, there has been accumulating evidence to suggest a strong communication between the gut and central nervous system, with microbes serving as modulators of several key neuroendocrine and neuroimmune pathways. Of note, recent work has suggested a strong link between certain characteristic microbiota and the presence of mood disorders or mood-related behavior, mostly in animal models. Of course, there exists the challenge of extending a perceived emotional response in an animal to that in a human, which frequently calls for heightened scrutiny. However, the utilization of animal models as a means of approximating behavioral changes in humans has been well-established through the use of verified behavioral testing, particularly for the development of anxiety-like and depression-like behaviors, and has been historically used by the United States pharmaceutical industry as a means of assessing the behavioral toxicology of drugs.[16] For a concise review of behavioral proxies used in rat models, see Belovicova et al.

Several alterations in the microbiome of animals have notably resulted in alterations in observed behavior that warrant further investigation. An example of increasing interest in the lay press is the potential for recurrent or extended antibiotic use to influence and perhaps alter mood. In fact, several animal studies using antibiotic-depleted gut microbe models have shown that antibiotic-treated animals display altered behavior, such as increased locomotor activity and altered anxiety and social interactions, compared to animals with an intact gut microbiome. Such behavioral changes are accompanied by corresponding physiologic changes in striatal synaptic proteins and levels of brain-derived neurotrophic factor in the amygdala.[17–20] This work has been supported by a notable human study as well; among a birth cohort of 871 European children and mothers, antibiotic use during the first year of life corresponded with the development of behavioral difficulties and depressive symptoms later in life.[21] Additional work is just beginning to elucidate the role of bacterial metabolites in modulating mood; MacFabe and colleagues, for example, have found that short-chain fatty acids (SCFAs) when introduced into pregnant or newborn mice, are capable of inducing long term behavioral changes in anxiety-like behavior, locomotion, and the startle reflex.[22] Finally, any discussion of the microbiome would be incomplete without mention of the explosive interest in fecal transplant as a means of pathogenesis or treatment. The fecal transplant hypothesis, though still highly controversial, is not without merit; germ-free rats that receive fecal transplants from depressed patients exhibit depression-like behavior, namely anhedonia, as well as alterations in tryptophan metabolism, while probiotic treatment with *Lactobacillus casei DG* has been shown to reverse this phenotype.[23,24] In short, while it is not possible given the current literature to definitively correlate certain taxonomic groupings with mood changes or disorders in humans, the accumulating evidence is strong and suggests a critical role for the gut microbiome in modulating mood, and therefore in the etiology and possibly treatment of mood disorders.

3.2 Major depressive disorder and microbiome dysregulation

While there are a growing number of human studies in the field, they have largely focused on physically and cognitively debilitating neuropsychiatric disorders, such as autism spectrum disorder (ASD), schizophrenia, and Parkinson's Disease, rather than variance in more "typical" well-being, such as depressive symptoms. This section seeks to present the current key findings, questions, and assumptions about the role of the gut-brain-axis in the pathogenesis, and potentially treatment, of major depressive disorder (MDD).

MDD is defined by the DSM-V as "discrete episodes of at least 2 weeks' duration (although most episodes last considerably longer) involving clear-cut changes in affect, cognition, and neurovegetative functions and interepisode remissions. A diagnosis based on a single episode is possible, although the disorder is a recurrent one in the majority of cases".[13] This is to be distinguished from dysthymia, a new diagnosis in the DSM-V, which is defined as a "more chronic form of depression," or more specifically "when the mood disturbance continues for at least 2 years in adults or 1 year in children."[13] We must acknowledge that several human studies have found conflicting associations between the gut microbiome and mood, focused mostly on MDD. For example, in a study of 79 non-randomly selected participants with either active MDD ($n=29$), responding MDD ($n=17$), or healthy controls ($n=30$), participants with active MDD had significantly increased fecal bacterial alpha diversity, a measure of the mean species diversity of the GI tract.[25] However, the authors acknowledge that many of their patients used atypical antipsychotics, which have been shown to increase the relative abundance of Firmicutes and decrease that of Bacteroidetes in the cecum and thus stands as a major limitation and potential confounder to this study.[25–27] In contrast, fecal samples from 34 depressed and 33 healthy matched controls, selected non-randomly from inpatient and outpatient psychiatry clinics, showed decreased diversity in participants with depression.[24] Also, a small study on patients with abdominal symptoms (n=17) that lacked a control group showed that heterologous fecal transplantation improved ratings on the Hamilton Rating Scale for Depression (HAM-D) and subscale of sleep-related terms, Hamilton Rating Scale for Anxiety (HAM-A) and Quick Inventory for Depressive Symptoms (QIDS).[28] Interestingly, the same study reported that at baseline, individuals with a HAM-D ≥ 8 had a Shannon index indicating lower microbial diversity, which improved with fecal transplantation and supports the hypothesis that decreased alpha diversity may be associated with depression.[28] Thus, mean diversity may not be a useful indicator of microbial dysbiosis; rather, specific shifts in the relative ratios of taxa, the metabolites they produce, or the bacteria they outcompete, may be more clinically useful. For example, in a recent, large Dutch population-based cohort ($n=1054$) including 80 individuals with MDD (40 undergoing antidepressant treatment) and 70 healthy controls, which takes into account confounding variables such as antidepressant treatment, participants with MDD showed depletion of specific gut genera, namely *Dialister*, anaerobic Gram-negative rods, and

Coprococcus, anaerobic butyric acid-producing cocci.[29] Similar studies support these findings, with patients with MDD also showing increased proportions of bacteria from the genera *Bacteroidetes*, *Protobacteria*, and *Actinobacteria*, and fewer *Firmicutes* and *Faecalibacterium*.[25] Interestingly, increased levels of butyrate-producing bacteria, namely *Faecalibacterium* and *Coprococcus*, have also been associated with higher quality of life indicators, as quantified by the RAND-36 health-related quality of life survey.[29] Butyrate is a short-chain fatty acid (SCFA) that strengthens the intestinal epithelium and reduces inflammation, and has been shown to be decreased in both MDD and inflammatory bowel disease.[29] Assessment of patients with Generalized Anxiety Disorder (GAD) also reveals a characteristic microbiome shift with some notable overlap with that found in patients with MDD. For example, patients with GAD were found to have considerable decreases in five genera, those being *Faecalibacterium*, *Eubacterium rectale*, *Lachnospira*, *Butyricicoccus*, and *Sutterella*.[30] Given the high comorbidity between GAD and MDD, this finding is of particular interest.

On a molecular level, one potential mechanism by which commensal bacteria, namely *Coprococcus comes* and *Coprococcus catus* may improve mood is via increased levels of 3,4-dihydroxyphenylacetic acid (DOPAC), one of the three metabolites of dopamine.[29] In their manuscript, Valles-Colomer and colleagues clearly state that covariation does not imply that *Coprococcus* species are responsible for directly synthesizing DOPAC; however, the genus was positively correlated with both mental and physical RAND scores, as well as increased levels of DOPAC, which suggests a potential link between positive quality of life and Coprococcus beyond the traditionally studied production of butyrate.[29] Work has long-postulated a crucial role for dopamine in the etiology of depressive symptoms, as deficits in mesolimbic dopaminergic pathways are thought to be responsible for the anhedonia frequently experienced in depression. It also cannot be ignored that many of the aforementioned neuropsychiatric disorders, namely Parkinson's Disease and schizophrenia, also result from deficits in dopaminergic pathways.[31]

It is also crucial to address the rapidly increasing interest in probiotic treatment for mood disorders. The current theory stands that consumption of probiotics attenuates the hypothalamic-pituitary-adrenal (HPA) axis hyperactivation, thereby decreasing pro-stress and pro-inflammatory secretion of adrenocorticotropic hormone (ACTH), corticosterone, adrenaline, and noradrenaline, with the most frequent probiotic administered being

Lactobacillus casei.[32] This is particularly impactful given that patients with MDD have been shown to exhibit HPA axis hyperactivity, including high circulating levels of corticotropin-releasing hormone in the cerebrospinal fluid.[33] Additionally, the evidence to suggest a potential therapeutic link is increasing, with no serious adverse reactions recorded. For example, a randomized controlled trial on 40 symptom-free young adults showed that the consumption of a multispecies probiotic containing *Bifidobacterium bifidum*, *Bifidobacterium lactis*, *Lactobacillus acidophilus*, *Lactobacillus brevis*, *Lactobacillus casei*, *Lactobacillus salivarius*, and *Lactococcus lactis* reduced cognitive reactivity, a behavioral trait considered a risk factor for depression.[34] Similarly, another randomized controlled parallel-arm study with two control groups on 46 healthy women showed that the consumption of probiotics was associated with brain activity in areas related to emotional processing.[35] Two meta-analyses showed that probiotics reduced depressive symptoms in participants with major depressive disorder[36] but not in healthy participants.[37] For excellent recent reviews discussing the relationship between the gut microbiota and depression and the gut microbiota and mental health more generally, see Refs. [6,38], respectively.

3.3 Intestinal inflammation and mood changes

Interestingly, there is mounting evidence to suggest a link between inflammation, mood dysregulation, and microbiome shifts. For example, certain bacterial endotoxins and antigens, such as lipopolysaccharide (LPS), bacterial lipoprotein, flagellin, and CpG oligodeoxylucleotide can stimulate the release of pro-inflammatory cytokines which both increase permeability of the intestinal mucosa as well as the blood-brain barrier.[39–41] The induction of a low-grade, pro-inflammatory intestinal mucosa has been observed in patients with depression.[42] One popular hypothesis linking inflammation to depression is that of decreased tryptophan metabolism; inflammation has been shown to induce the tryptophan metabolizing enzyme indoleamine 2,3-dioxygenase and thus may lead to lower plasma tryptophan levels and diminishing serotonin activity in the brain.[43,44] The role of SCFAs within the gut in modulating human mood is less clear; while introduction of SCFA-producing bacteria induced an anxiety-like state in murine models, the reduction of such bacteria in humans has been associated with inflammatory bowel disease and autoimmunity.[22,45] However, one recent study on young Americans comparing 43 participants with major MDD and 47 healthy controls predicted that functional differences in the gut microbiome between groups included reduced short-chain fatty acid production pathways in the MDD group.[46]

It is also important to acknowledge that the normal intestinal immune response is largely influenced by the presence of commensal bacteria within the developing gut. The colonized gut uses the presence of toll-like receptors (TLRs) to identify pathogen-associated molecular patterns (PAMPs) characteristic of pathogenic bacteria to mount an immune response against them.[47] In murine models, however, the germ-free environment has produced animals with diminished or absent profiles of certain TLRs, reduced IgA secretion, and fewer and smaller Peyer's patches in the intestinal mucosa.[48] These animals, in turn, produce a diminished immune response to lipopolysaccharide (LPS) stimulation, the normal response to which can be rescued by gut repopulation.[49,50] Interestingly, additional work has shown germ-free mice to have decreased levels of regulatory T cells with reduced anti-inflammatory properties, which further supports the hypothesis that gut dysregulation may be associated with immune dysfunction.[51] Increasing work has also proposed an immune-mediated etiology to MDD, with the gut microbiome playing a central role. For example, one study attempted to identify potential mediators between depression and the microbiome by recruiting a convenience sample of 112 patients with MDD and 28 healthy volunteers from a local outpatient clinic and measuring immunoglobulin levels, and found increased IgA and IgM responses to common gut commensals in depressed participants.[52] Since stress-induced inflammation has long been associated with MDD, it is therefore possible that there is an autoimmune component to the pathogenesis of MDD in certain individuals.[53]

3.4 The microbiome and neurotransmitter synthesis

There is also an increasingly clear role for gut microbes in the production and reception of neurotransmitters and hormones. For example, it has been estimated that approximately 90% of endogenous serotonin levels are produced in the gut, the alteration of which can have profound effects on an individual's mood.[54] Preceding work supports this, with a 2011 study postulating that commensal bacteria such as *Bifidobacteria infantis* can modulate tryptophan metabolism and in turn 5-HT levels.[55] Looking beyond MDD, additional work has demonstrated a possible role for norepinephrine/noradrenaline (NE) in inter-kingdom communication; despite bacteria having no known adrenergic receptors, it is possible that NE acts as a bacterial siderophore, resulting in the overexpression of enterobactin and the iron-chelating mechanism of *E. coli*, thereby increasing the bacterial growth rate.[56] Among a convenience sample of 38 healthy human males, fecal transplantation led to changes in the recipient's fecal microbial composition and increased plasma levels of GABA, though these changes were not maintained long-term.[57] In short, while the links are not yet fully understood, the evidence is strong to suggest that the composition of one's gut may very well alter mood through modulation of neurotransmitter levels.

3.5 Conclusion

Current research associating the gut microbiome and behavior is still in its infancy as cause and effect is inconclusive; however as research grows exponentially, the link becomes stronger. Evidence continues to grow showing various possible mechanisms causing dysregulation to the microbiota environment impacting mood and its encompassing range of disorders. Further animal and human studies are warranted to investigate mechanistic processes and potential treatments.

4. Microbiome and cognition

Cognition refers to the mental process of acquiring knowledge and comprehension through thought and experience. Neurocognitive functions encompass six key domains, which include perception, social cognition, attention/concentration, learning and memory, executive function, and language abilities.[13] While cognitive decline is a natural process in normal aging, the impairment of cognition that is severe enough to interfere with activities of daily living is considered to be a disease process. For example, dementia is a disorder characterized by impairment of the memory domain with at least one of the other cognitive function domains. Similarly, patients with Alzheimer's disease also have memory and executive function impairment. Experiments on the microbiome through diet and manipulation of specific microbes suggest direct effects on cognition, including learning, memory, and disease progression.

Animal models have provided a robust body of evidence linking the gut microbiome with cognitive functioning. In animal models, germ-free mice show memory impairment and reduced hippocampal function compared to normal mice.[58] Antibiotic-treated mice which are pseudo germ-free show decreased alpha diversity and impaired memory.[59] Similarly, mice consuming specific probiotic species (*Bifidobacterium longum* 1714) performed better on learning and memory tasks compared to controls but this same effect was not seen in mice consuming *Bifidobacterium breve* 1205.[60] In a murine model of Alzheimer's disease, specific gut-microbial derived metabolites, urolithins, inhibit the development of amyloid fibrils.[61] Amyloid fibril plaques are found in brain tissue of Alzheimer patients. The findings of Yuan et al. imply that specific microbial metabolites may contribute to prevention of neurodegenerative diseases.

Some similar associations have been found in limited human studies, but large gaps remain, especially studies exploring the relationship between gut microbiome and typical cognitive functioning in adults. A few studies have focused on children. For example, among 89 infants selected from two prospective cohorts, alpha diversity was inversely associated with cognitive function.[62] Another study of 39 infants from a prospective cohort found that alpha diversity was associated with functional connectivity between certain brain regions that are related to cognitive outcomes at 2 years of age, providing a potential mechanistic pathway between the gut and cognition during development.[12] While these studies are intriguing, given the small sample sizes and focus on infants and toddlers, we are limited as to what we can infer regarding cognitive functioning over the life cycle.

Adult studies are largely focused on abnormal cognitive functioning. In 25 participants with Alzheimer's disease, alterations in the gut composition were seen when compared to 88 non-demented age-and sex-matched controls.[63] Based on a large multi-center study, researchers found that bacterially-produced bile acids are associated with cognitive decline in participants with Alzheimer's disease when compared to cognitively normal individuals, though there was no accounting for earlier cognitive function.[64] Based on 25 non-randomly selected Alzheimer's participants, shifts in the microbiome towards increased *Bacteroides* was associated with a greater amyloid burden in the brain as compared to 25 age- and sex-matched controls. [63] Among 40 non-randomly selected cognitively impaired adults with brain amyloidosis, there was a higher abundance of pro-inflammatory Escherichia/Shigella and lower abundance of anti-inflammatory Eubacterium rectale in stool as well as higher levels of pro-inflammatory cytokines compared to 33 participants without brain amyloidosis.[65] In a cross-sectional study comparing demented ($n=34$) versus non-demented ($n=94$) individuals, the percentage of *Bacteroides* was significantly lower in the dementia group.[66] That said, in one small study of 43 community-dwelling healthy older adults, different phyla such as Bacteroidetes, Firmicutes, Proteobacteria, Verrucomicrobia were associated with cognitive test performance, including attention and executive functioning, though no covariates were accounted for during statistical analysis, which likely biased the findings.[67] One large cohort study found a negative association between reaction time and alpha diversity as well as Burkholderiales in a subset of 561 participants, after accounting for frailty and dietary factors, and excluding those taking antibiotics or proton pump inhibitors.[68]

Early studies show promising results associating pro-inflammatory gut microbiomes with neurocognitive diseases, such as dementia and Alzheimer's disease. Randomized prospective studies are needed to establish causation relationships between microbiomes and cognitive disorders. These studies will provide valuable insights on the development of alternative diagnostic surveillance and therapeutic interventions targeting gut microbiomes in neurocognitive disorders.

5. Mechanism of action

The mechanisms by which gut microbes can influence the brain include the production of neuroactive and metabolically active substances, vagal afferents, short chain fatty acid levels, immune-mediated responses, and modulation of blood-brain barrier (BBB) integrity.[69–73] Neuroactive substances such as norepinephrine (NE), dopamine (DA), serotonin (5-HT), γ-aminobutyric acid (GABA), histamine, short chain fatty acids (SCFAs), and brain-derived neurotrophic factor (BDNF) are produced by gut commensal bacteria[17,19,73–75] and these substances, in and of themselves, play important roles in human behavior and cognition such as reward, memory, learning, and attention.[73] Among bacteria, these molecules generally have important roles in microbial community dynamics. For example, bacteria use norepinephrine as a quorum sensing molecule,[76] and GABA secretion allows for regulation of intracellular pH.[77] In the presence of dopamine and norepinephrine, *Escherichia coli* O157:H7 exhibit increased growth rate, motility, biofilm formation, and virulence[78] and certain bacteria can use GABA as their sole carbon and nitrogen source.[79] In fact, it has been proposed that many cell-cell signaling molecules operating across immune, metabolic, and nervous system pathways have been acquired in animals through horizontal gene transfer from bacteria,[80] although metagenomic data suggests these compounds are rare in plasmids.[29]

Researchers use probiotics and other interventions to identify potential mechanisms, and some studies have revealed associations with changes in the brain and/or behavior. For example, animals fed probiotic bacteria show changes in brain GABA levels and reduced anxiety-like behavior, both of which are dependent on the vagus nerve.[81] Mice whose gut microbiota were altered through exercise and diet demonstrated different patterns in anxiety and cognition.[82] Germ-free mice fed SCFAs such as butyrate have increased BBB integrity compared to conventional mice.[70] In an autistic mouse model, butyrate improved social behavior and altered expression of behavior-related genes in the prefrontal cortex.[83] In humans, depression has been linked to increased inflammatory responses[52] and cognitive differences have been associated with increased alpha diversity and regional brain interconnectivity.[12]

6. Conclusions

It should come as little surprise that the interaction between the gut microbiome and psychocognitive traits is extremely complex and not yet well understood. In this chapter, we have presented evidence from the literature that supports a dynamic bi-directional interaction between personality, mood, and cognition, and gut microbial composition. The gut should be thought of as not simply a tube through which nutrients pass, but rather a selectively permeable and adaptable interface through which the communities that reside within us can, through mechanisms we are just beginning to understand, modulate our central nervous system. As we have summarized, this may occur through variations in bacterial metabolites, neurotransmitter synthesis or decreases thereof, or complex interactions that occur as a result of shifting ratios between microbial taxa, or perhaps alpha diversity itself.

The study of the microbiome in the context of behaviorism holds immense potential to elucidate not only the role of bacteria in shaping humanity, but also their role in devastating human diseases such as Alzheimer's dementia. Thus, as we look towards the future, and the evidence surrounding fecal transplants and probiotics in the treatment of psychiatric disorders continues to accumulate, it is now our belief that any discussion of aberrant behavioral development will soon include an analysis and discussion of an individual's microbiome. Ultimately, we are confident that the use of "alternative" therapies such as probiotics being offered in the treatment of irritable bowel disease, for example, will expand to psychiatric illness and perhaps even cognitive diseases as well.

References

1. Jung CG. *Psychological Types. Collected Works of C.G. Jung.* vol 6. Princeton University Press; 1971. ISBN 0-691-09770-4.
2. Cattell RB. *The Scientific Analysis of Personality.* Baltimore: Penguin Books; 1965.
3. Costa PT, McCrae RR. The five-factor model of personality and its relevance to personality disorders. *J Pers Disord.* 1992;6(4):343–359. https://doi.org/10.1521/pedi.1992.6.4.343.
4. Goldberg LR. The structure of phenotypic personality traits. *Am Psychol.* 1993; 48(1):26–34. https://doi.org/10.1037//0003-066x.48.1.26.
5. Markett S, Montag C, Reuter M. Network neuroscience and personality. *Personal Neurosci.* 2018;1, e14. Published 2018 Aug 10 https://doi.org/10.1017/pen.2018.12.
6. Dinan TG, Cryan JF. Gut microbes and depression: still waiting for Godot. *Brain Behav Immun.* 2019;79:1–2. https://doi.org/10.1016/j.bbi.2019.02.007.
7. Buffington SA, Di Prisco GV, Auchtung TA, Ajami NJ, Petrosino JFM Costa-Mattioli. Microbial reconstitution reverses maternal diet-induced social and synaptic deficits in offspring. *Cell.* 2016;165(7):1762–1775. https://doi.org/10.1016/j.cell.2016.06.001.
8. Sanguinetti E, Guzzardi MA, Tripodi M, et al. Microbiota signatures relating to reduced memory and exploratory behaviour in the offspring of overweight mothers in a murine model. *Sci Rep.* 2019;9:12609. https://doi.org/10.1038/s41598-019-48090-8.
9. Kim HN, Yun Y, Ryu S, et al. Correlation between gut microbiota and personality in adults: a cross-sectional study. *Brain Behav Immun.* 2018;69:374–385. https://doi.org/10.1016/j.bbi.2017.12.012.
10. Christian LM, Galley JD, Hade EM, Schoppe-Sullivan S, Kamp Dush C, Bailey MT. Gut microbiome composition is associated with temperament during early childhood. *Brain Behav Immun.* 2015;45:118–127. https://doi.org/10.1016/j.bbi.2014.10.018.
11. Aatsinki AK, Lahti L, Uusitupa HM, et al. Gut microbiota composition is associated with temperament traits in infants. *Brain Behav Immun.* 2019;80:849–858. https://doi.org/10.1016/j.bbi.2019.05.035.
12. Gao W, Salzwedel AP, Carlson AL, et al. Gut microbiome and brain functional connectivity in infants-a preliminary study focusing on the amygdala. *Psychopharmacology (Berl).* 2019;236(5):1641–1651. https://doi.org/10.1007/s00213-018-5161-8.
13. American Psychiatric Association. In: *Diagnostic and Statistical Manual of Mental Disorders.* 5th ed. Washington, DC: American Psychiatric Association; 2013.
14. Carlat DJ. *The Psychiatric Interview: A Practical Guide.* 2nd ed. Philadelphia: Lippincott Williams and Wilkins; 2005:122–143. Mental status examination.
15. Lee C, Doo E, Choi JM, et al. The increased level of depression and anxiety in irritable bowel syndrome patients compared with healthy controls: systematic review and meta-analysis. *J Neurogastroenterol Motil.* 2017;23(3):349–362.
16. Belovicova K, Bogi E, Csatlosova K, Dubovicky M. Animal tests for anxiety-like and depression-like behavior in rats. *Interdiscip Toxicol.* 2017;10(1):40–43.
17. Arentsen T, Raith H, Qian Y, Forssberg H, Diaz Heijtz R. Host microbiota modulates development of social preference in mice. *Microb Ecol Health Dis.* 2015; 26:29719.
18. Desbonnet L, Clarke G, Shanahan F, Dinan TG, Cryan JF. Microbiota is essential for social development in the mouse. *Mol Psychiatry.* 2014;19(2):146–148.
19. Diaz Heijtz R, Wang S, Anuar F, et al. Normal gut microbiota modulates brain development and behavior. *Proc Natl Acad Sci U S A.* 2011;108(7):3047–3052.
20. Neufeld KM, Kang N, Bienenstock J, Foster JA. Reduced anxiety-like behavior and central neurochemical change in germ-free mice. *Neurogastroenterol Motil.* 2011; 23(3):255–264. e119.

21. Slykerman RF, Thompson J, Waldie KE, Murphy R, Wall C, Mitchell EA. Antibiotics in the first year of life and subsequent neurocognitive outcomes. *Acta Paediatr.* 2017; 106(1):87–94.
22. MacFabe DF. Enteric short-chain fatty acids: microbial messengers of metabolism, mitochondria, and mind: implications in autism spectrum disorders. *Microb Ecol Health Dis.* 2015;26:28177.
23. Guida F, Turco F, Iannotta M, et al. Antibiotic-induced microbiota perturbation causes gut endocannabinoidome changes, hippocampal neuroglial reorganization and depression in mice. *Brain Behav Immun.* 2018;67:230–245.
24. Kelly JR, Borre Y, O' Brien C, et al. Transferring the blues: depression-associated gut microbiota induces neurobehavioural changes in the rat. *J Psychiatr Res.* 2016;82: 109–118.
25. Jiang H, Ling Z, Zhang Y, et al. Altered fecal microbiota composition in patients with major depressive disorder. *Brain Behav Immun.* 2015;48:186–194.
26. Davey KJ, Cotter PD, O'Sullivan O, et al. Antipsychotics and the gut microbiome: olanzapine-induced metabolic dysfunction is attenuated by antibiotic administration in the rat. *Transl Psychiatry.* 2013;3, e309.
27. Davey KJ, O'Mahony SM, Schellekens H, et al. Gender-dependent consequences of chronic olanzapine in the rat: effects on body weight, inflammatory, metabolic and microbiota parameters. *Psychopharmacology (Berl).* 2012;221:155–169.
28. Kurokawa S, Kishimoto T, Mizuno S, et al. The effect of fecal microbiota transplantation on psychiatric symptoms among patients with irritable bowel syndrome, functional diarrhea and functional constipation: an open-label observational study. *J Affect Disord.* 2018;235:506–512.
29. Valles-Colomer M, Falony G, Darzi Y, et al. The neuroactive potential of the human gut microbiota in quality of life and depression. *Nat Microbiol.* 2019;4(4):623–632.
30. Jiang HY, Zhang X, Yu ZH, et al. Altered gut microbiota profile in patients with generalized anxiety disorder. *J Psychiatr Res.* 2018;104:130–136.
31. Belujon P, Grace AA. Dopamine system dysregulation in major depressive disorders. *Int J Neuropsychopharmacol.* 2017;20(12):1036–1046.
32. Wallace CJK, Milev R. The effects of probiotics on depressive symptoms in humans: a systematic review [published correction appears in Ann Gen Psychiatry. 2017 Mar 7; 16: 18]. *Ann Gen Psychiatry.* 2017;16:14.
33. Varghese FP, Brown ES. The hypothalamic-pituitary-adrenal axis in major depressive disorder: a brief primer for primary care physicians. *Prim Care Companion J Clin Psychiatry.* 2001;3(4):151–155.
34. Steenbergen L, Sellaro R, van Hemert S, Bosch JA, Colzato LS. A randomized controlled trial to test the effect of multispecies probiotics on cognitive reactivity to sad mood. *Brain Behav Immun.* 2015;48:258–264.
35. Tillisch K, Labus J, Kilpatrick L, et al. Consumption of fermented milk product with probiotic modulates brain activity. *Gastroenterology.* 2013;144(7):1394–1401. e14014.
36. Huang R, Wang K, Hu J. Effect of probiotics on depression: a systematic review and meta-analysis of randomized controlled trials. *Nutrients.* 2016;8(8):483.
37. Ng QX, Peters C, Ho CYX, Lim DY, Yeo WS. The neuroactive potential of the human gut microbiota in quality of life and depression. *Nat Microbiol.* 2019;4(4):623–632.
38. Du Toit A. The gut microbiome and mental health. *Nat Rev Microbiol.* 2019;17(4):196.
39. Huang TT, Lai JB, Du YL, Xu Y, Ruan LM, Hu SH. Current understanding of gut microbiota in mood disorders: an update of human studies. *Front Genet.* 2019;10:98.
40. Petra AI, Panagiotidou S, Hatziagelaki E, Stewart JM, Conti P, Theoharides TC. Gut-microbiota-brain axis and its effect on neuropsychiatric disorders with suspected immune dysregulation. *Clin Ther.* 2015;37(5):984–995.

41. Sampson TR, Mazmanian SK. Control of brain development, function, and behavior by the microbiome. *Cell Host Microbe.* 2015;17(5):565–576.
42. Kelly JR, Kennedy PJ, Cryan JF, Dinan TG, Clarke G, Hyland NP. Breaking down the barriers: the gut microbiome, intestinal permeability and stress-related psychiatric disorders. *Front Cell Neurosci.* 2015;9:392.
43. Cowen PJ, Browning M. What has serotonin to do with depression? *World Psychiatry.* 2015;14(2):158–160.
44. Wichers MC, Koek GH, Robaeys G, Verkerk R, Scharpé S, Maes M. IDO and interferon-alpha-induced depressive symptoms: a shift in hypothesis from tryptophan depletion to neurotoxicity. *Mol Psychiatry.* 2005;10(6):538–544.
45. Clemente JC, Manasson J, Scher JU. The role of the gut microbiome in systemic inflammatory disease. *BMJ.* 2018;360, j5145.
46. Liu RT, Rowan-Nash AD, Sheehan AE, et al. Reductions in anti-inflammatory gut bacteria are associated with depression in a sample of young adults. *Brain Behav Immun.* 2020;88:308–324. https://doi.org/10.1016/j.bbi.2020.03.026.
47. Akira S, Hemmi H. Recognition of pathogen-associated molecular patterns by TLR family. *Immunol Lett.* 2003;85(2):85–95.
48. Luczynski P, McVey Neufeld KA, Oriach CS, Clarke G, Dinan TG, Cryan JF. Growing up in a bubble: using germ-free animals to assess the influence of the gut microbiota on brain and behavior. *Int J Neuropsychopharmacol.* 2016;19(8), pyw020.
49. Clarke G, Grenham S, Scully P, et al. The microbiome-gut-brain axis during early life regulates the hippocampal serotonergic system in a sex-dependent manner. *Mol Psychiatry.* 2013;18:666–673.
50. Umesaki Y, Okada Y, Matsumoto S, Imaoka A, Setoyama H. Segmented filamentous bacteria are indigenous intestinal bacteria that activate intraepithelial lymphocytes and induce MHC class II molecules and fucosyl asialo GM1 glycolipids on the small intestinal epithelial cells in the ex-germ-free mouse. *Microbiol Immunol.* 1995;39:555–562.
51. Ostman S, Rask C, Wold AE, Hultkrantz S, Telemo E. Impaired regulatory T cell function in germ-free mice. *Eur J Immunol.* 2006;36(9):2336–2346.
52. Maes M, Kubera M, Leunis JC, Berk M. Increased IgA and IgM responses against gut commensals in chronic depression: further evidence for increased bacterial translocation or leaky gut. *J Affect Disord.* 2012;141(1):55–62.
53. Peirce JM, Alviña K. The role of inflammation and the gut microbiome in depression and anxiety. *J Neurosci Res.* 2019;97(10):1223–1241.
54. Yano JM, Yu K, Donaldson GP, et al. Indigenous bacteria from the gut microbiota regulate host serotonin biosynthesis [published correction appears in Cell. 2015 Sep 24; 163:258]. *Cell.* 2015;161(2):264–276.
55. Desbonnet L, Garrett L, Clarke G, Bienenstock J, Dinan TG. The probiotic Bifidobacteria infantis: an assessment of potential antidepressant properties in the rat. *J Psychiatr Res.* 2008;43(2):164–174.
56. Lazar V, Ditu LM, Pircalabioru GG, et al. Aspects of gut microbiota and immune system interactions in infectious diseases, immunopathology, and cancer. *Front Immunol.* 2018; 9(15):1830.
57. Kootte RS, Levin E, Salojarvi J, et al. Improvement of insulin sensitivity after lean donor feces in metabolic syndrome is driven by baseline intestinal microbiota composition. *Cell Metab.* 2017;26:611–619. e6.
58. Gareau MG, Wine E, Rodrigues DM, et al. Bacterial infection causes stress-induced memory dysfunction in mice. *Gut.* 2011;60(3):307–317. https://doi.org/10.1136/gut.2009.202515.
59. Zhan G, Yang N, Li S, et al. Abnormal gut microbiota composition contributes to cognitive dysfunction in SAMP8 mice. *Aging (Albany NY).* 2018;10(6):1257–1267. https://doi.org/10.18632/aging.101464.

60. Savignac HM, Tramullas M, Kiely B, Dinan TG, Cryan JF. Bifidobacteria modulate cognitive processes in an anxious mouse strain. *Behav Brain Res*. 2015;287:59–72. https://doi.org/10.1016/j.bbr.2015.02.044.
61. Yuan T, Ma H, Liu W, et al. Pomegranate's neuroprotective effects against Alzheimer's disease are mediated by urolithins, its ellagitannin-gut microbial derived metabolites. *ACS Chem Neurosci*. 2016;7(1):26–33. https://doi.org/10.1021/acschemneuro.5b00260.
62. Carlson AL, Xia K, Azcarate-Peril MA, et al. Infant gut microbiome associated with cognitive development. *Biol Psychiatry*. 2018;83(2):148–159. https://doi.org/10.1016/j.biopsych.2017.06.021.
63. Vogt NM, Kerby RL, Dill-McFarland KA, et al. Gut microbiome alterations in Alzheimer's disease. *Sci Rep*. 2017;7(1):13537. https://doi.org/10.1038/s41598-017-13601-y.
64. MahmoudianDehkordi S, Arnold M, Nho K, et al. Altered bile acid profile associates with cognitive impairment in Alzheimer's disease—an emerging role for gut microbiome. *Alzheimers Dement*. 2019;15(1):76–92. https://doi.org/10.1016/j.jalz.2018.07.217.
65. Cattaneo A, Cattane N, Galluzzi S, et al. Association of brain amyloidosis with pro-inflammatory gut bacterial taxa and peripheral inflammation markers in cognitively impaired elderly. *Neurobiol Aging*. 2017;49:60–68. https://doi.org/10.1016/j.neurobiolaging.2016.08.019.
66. Saji N, Niida S, Murotani K, et al. Analysis of the relationship between the gut microbiome and dementia: a cross-sectional study conducted in Japan. *Sci Rep*. 2019; 9(1):1008. https://doi.org/10.1038/s41598-018-38218-7.
67. Manderino L, Carroll I, Azcarate-Peril MA, et al. Preliminary evidence for an association between the composition of the gut microbiome and cognitive function in neurologically healthy older adults. *J Int Neuropsychol Soc*. 2017;23(8):700–705. https://doi.org/10.1017/S1355617717000492.
68. Verdi S, Jackson MA, Beaumont M, et al. An investigation into physical frailty as a link between the gut microbiome and cognitive health. *Front Aging Neurosci*. 2018;10:398. https://doi.org/10.3389/fnagi.2018.00398.
69. Bonaz B, Bazin T, Pellissier S. The vagus nerve at the interface of the microbiota-gut-brain axis. *Front Neurosci*. 2018;12:49.
70. Braniste V, Al-Asmakh M, Kowal C, et al. The gut microbiota influences blood-brain barrier permeability in mice. *Sci Transl Med*. 2014;6(263):263ra158.
71. Dalile B, Van Oudenhove L, Vervliet B, Verbeke K. The role of short-chain fatty acids in microbiota–gut–brain communication. *Nat Rev Gastroenterol Hepatol*. 2019; 16(8):461–478.
72. Logsdon A, Erickson M, Rhea E, Salameh T, Banks W. Gut reactions: how the blood–brain barrier connects the microbiome and the brain. *Exp Biol Med*. 2017; 243(2):159–165.
73. Strandwitz P. Neurotransmitter modulation by the gut microbiota. *Brain Res*. 2018;1693:128–133.
74. Lyte M. The role of microbial endocrinology in infectious disease. *J Endocrinol*. 1993; 137(3):343–345. https://doi.org/10.1677/joe.0.1370343.
75. Asano Y, Hiramoto T, Nishino R, et al. Critical role of gut microbiota in the production of biologically active, free catecholamines in the gut lumen of mice. *Am J Physiol Gastrointest Liver Physiol*. 2012;303(11):G1288–G1295. https://doi.org/10.1152/ajpgi.00341.2012.
76. Sperandio V, Torres AG, Jarvis B, Nataro JP, Kaper JB. Bacteria-host communication: the language of hormones. *Proc Natl Acad Sci USA*. 2003;100(15):8951–8956. https://doi.org/10.1073/pnas.1537100100.

77. Feehily C, Karatzas KA. Role of glutamate metabolism in bacterial responses towards acid and other stresses. *J Appl Microbiol.* 2013;114(1):11–24. https://doi.org/10.1111/j.1365-2672.2012.05434.x.
78. Bansal T, Englert D, Lee J, Hegde M, Wood TK, Jayaraman A. Differential effects of epinephrine, norepinephrine, and indole on Escherichia coli O157:H7 chemotaxis, colonization, and gene expression. *Infect Immun.* 2007;75(9):4597–4607. https://doi.org/10.1128/IAI.00630-07.
79. Dover S, Halpern YS. Utilization of -aminobutyric acid as the sole carbon and nitrogen source by Escherichia coli K-12 mutants. *J Bacteriol.* 1972;109(2):835–843. https://doi.org/10.1128/JB.109.2.835-843.1972.
80. Iyer LM, Aravind L, Coon SL, Klein DC, Koonin EV. Evolution of cell-cell signaling in animals: did late horizontal gene transfer from bacteria have a role? *Trends Genet.* 2004;20(7):292–299. https://doi.org/10.1016/j.tig.2004.05.007.
81. Bravo JA, Forsythe P, Chew MV, et al. Ingestion of Lactobacillus strain regulates emotional behavior and central GABA receptor expression in a mouse via the vagus nerve. *Proc Natl Acad Sci USA.* 2011;108(38):16050–16055. https://doi.org/10.1073/pnas.1102999108.
82. Kang SS, Jeraldo PR, Kurti A, et al. Diet and exercise orthogonally alter the gut microbiome and reveal independent associations with anxiety and cognition. *Mol Neurodegener.* 2014;9:36. https://doi.org/10.1186/1750-1326-9-36.
83. Kratsman N, Getselter D, Elliott E. Sodium butyrate attenuates social behavior deficits and modifies the transcription of inhibitory/excitatory genes in the frontal cortex of an autism model. *Neuropharmacology.* 2016;102:136–145. https://doi.org/10.1016/j.neuropharm.2015.11.003.

CHAPTER FIVE

Bioinformatics and machine learning in gastrointestinal microbiome research and clinical application

Lindsay M. Hopson[a,b,†], Stephanie S. Singleton[a,†], John A. David[d], Atin Basuchoudhary[e], Stefanie Prast-Nielsen[f], Pavel Klein[g], Sabyasachi Sen[a,c], and Raja Mazumder[a,b,*]

[a]Department of Biochemistry and Molecular Medicine, The George Washington University, Washington, DC, United States
[b]The McCormick Genomic and Proteomic Center, The George Washington University, Washington, DC, United States
[c]Department of Medicine, The George Washington University, Washington, DC, United States
[d]Department of Applied Mathematics, Virginia Military Institute, Lexington, VA, United States
[e]Department of Economics and Business, Virginia Military Institute, Lexington, VA, United States
[f]Center for Translational Microbiome Research (CTMR), Department of Microbiology, Tumor and Cell Biology, Karolinska Institutet, Stockholm, Sweden
[g]Mid-Atlantic Epilepsy and Sleep Center, Bethesda, MD, United States
*Corresponding author: e-mail address: mazumder@gwu.edu

Contents

† Authors contributed equally.

Progress in Molecular Biology and Translational Science, Volume 176
ISSN 1877-1173
https://doi.org/10.1016/bs.pmbts.2020.08.011

Abstract

The scientific community currently defines the human microbiome as all the bacteria, viruses, fungi, archaea, and eukaryotes that occupy the human body. When considering the variable locations, composition, diversity, and abundance of our microbial symbionts, the sheer volume of microorganisms reaches hundreds of trillions. With the onset of next generation sequencing (NGS), also known as high-throughput sequencing (HTS) technologies, the barriers to studying the human microbiome lowered significantly, making in-depth microbiome research accessible. Certain locations on the human body, such as the gastrointestinal, oral, nasal, and skin microbiomes have been heavily studied through community-focused projects like the Human Microbiome Project (HMP). In particular, the gastrointestinal microbiome (GM) has received significant attention due to links to neurological, immunological, and metabolic diseases, as well as cancer. Though HTS technologies allow deeper exploration of the GM, data informing the functional characteristics of microbiota and resulting effects on human function or disease are still sparse. This void is compounded by microbiome variability observed among humans through factors like genetics, environment, diet, metabolic activity, and even exercise; making GM research inherently difficult to study. This chapter describes an interdisciplinary approach to GM research with the goal of mitigating the hindrances of translating findings into a clinical setting. By applying tools and knowledge from microbiology, metagenomics, bioinformatics, machine learning, predictive modeling, and clinical study data from children with treatment-resistant epilepsy, we describe a proof-of-concept approach to clinical translation and precision application of GM research.

1. Introduction

The human microbiome has long been an understudied and poorly understood piece of the human-health puzzle, largely due to inadequate laboratory techniques and technology. Beginning in the late 1800s, Robert Koch was arguably one of the first scientists to demonstrate proper laboratory techniques for culturing bacteria,[1] setting the stage for microbiology and bacteriology for years to come. Other notable scientists include Theodour Escherich, who identified the common gastrointestinal bacteria *Escherichia coli*,[2] and Élie Metchnikoff, who discovered the concept of phagocytosis.[3] For many years, bacteria could only be studied through culturing, as only a small fraction of bacteria can be cultured from most environments.[4–7] Due to the viable but non-culturable (VBNC) state that bacteria can be present in,[8] there remained a huge gap in our understanding of the microbiome in general, until the advent of high-throughput sequencing technologies (HTS). With the relatively recent advancements in HTS, metagenomics research, and bioinformatics, researchers have been able to recognize the importance and influence that microbes have in overall human health, such as food digestion, vitamin production, metabolism, and fermentation.[9–11] Such knowledge was attainable with the advent of HTS and key initiatives, such as the Human Microbiome Project (HMP),[12] which launched more than a decade ago. The HMP sought to elucidate the entwined complexities of the human microbiome, health, and disease through a collaborative and interdisciplinary effort by the National Institutes of Health (NIH), The Broad Institute, Baylor College of Medicine, Washington University School of Medicine, J. Craig Venter Institute, and the Data Analysis and Coordination Center (DACC).[12] Human-associated microbial communities from healthy individuals were collected, characterized, and cataloged from several sites on the human body, including nasal passages, oral cavity, skin, gastrointestinal tract, and the urogenital tract. This collection of information captured the distribution and diversity observed across variable body sites.[12] The work of the HMP helped establish a baseline of the array of microbial communities that inhabit humans, along with three main concepts:

(1) *Microbial communities vary based on their locations in/on the human body and between individuals*
(2) *The human microbiome changes and diversifies over time*
(3) *Alterations and aberrations to the human microbiome have been associated with disease*

Since its inception, the HMP has become a world-renowned, benchmark resource for microbiome research. Though the initial HMP concluded in 2013, phase two, named the Integrated Human Microbiome Project (iHMP; https://hmpdacc.org/ihmp/),[13] aims to broaden its scope and breadth as a resource through a longitudinal, multi-omics approach. The iHMP focuses on delineating disease-specific cohorts (pregnancy and preterm birth, inflammatory bowel disease, and type 2 diabetes) shown to have measurable differences in microbiota compared to "healthy" individuals. This modernized approach utilizes improved sequencing technologies and platforms, as well as the implementation of more advanced methodologies for characterization, annotation, and outlining of functional characteristics.[14] The work of the HMP and iHMP has helped inform countless large-scale microbiome studies, many of which implement multi-omics, interdisciplinary, and crowd-sourced approaches, such as The American Gut Project (AGP; http://americangut.org),[15] the Earth Microbiome Project (EMP; http://www.earthmicrobiome.org/),[16] and the Human Food Project (HFP; http://humanfoodproject.com/), as well as academic research labs, consortiums, pharmaceutical companies, and start-up biotechnology labs. In addition, microbiome research has continued to gain importance and recognition through support such as National Microbiome Initiative.[17,18] Despite these efforts, much remains uncovered regarding the microbiome's biology, host interactions and associated responses, the impact of the presence, absence, and functionality of a specific organism or cluster of organisms, and the causes and effects of dysbiosis.[19–21]

In particular, the gastrointestinal microbiome (gut microbiome or GM) has received significant attention in recent years due to its distinct connections to metabolic, endocrine, immunological, and neurological activity and function.[22–25] It was once believed that the gut microbiome had approximately 10 times more microbial cells in the human gut than the entire human body, representing potentially 5000 different species and totaling nearly 100 trillion microbes.[26] Continued research in the field has allowed this theory to evolve, adjusting this statistic to reflect that the human gut microbiome contains at least as many cells as there are in the rest of our body (a ratio of about 1–1,3) and holds orders of magnitude more genes than the human genome[27]; making the gut microbiome extremely populated and diverse in both abundance and function. Much of the work in this research field has focused on characterizing commensal, symbiotic, and pathogenic bacteria (i.e., bacteriome), however, researchers are also exploring the vast array of archaea (archaeome), viruses (virome), fungi (mycobiome), and other unicellular or multicellular organisms like protists and helminths (eukaryome), in combination with their bacterial neighbors.[28–31] These organisms are collectively known as the *microbiota* and their collective genomes are referred to as the *microbiome*.[32]

It is well known within the scientific community that extensive microbial colonization is initiated at birth, with some preliminary studies suggesting the gut microbiome may begin establishing sometime during fetal development,[33] however, the latter statement is still a heavily debated topic. Studies have shown the neonatal microbiome to be influenced by delivery method (vaginal; cesarean section (C-section)), feeding mechanism (breast milk; formula), as well as the surrounding environment.[34] For example, it

measures of accuracy are computed for each approach, a simple first number (as seen in Fig. 2) is simply the number of correct out of sample predictions. From there, one can further investigate some of the more promising methods using a variety of metrics.

We use the area under the curve (AUC) of the Receiver Operator Characteristic (ROC), a standard evaluation metric for binary classification problems, to further evaluate the predictive value of the ensemble bagging algorithm in the test set (see Fig. 3). The idea of the ROC originated when the British started using radar to identify enemy aircraft in World War II. As they increased radar sensitivity, they began to pick out more enemy aircraft, as well as more flocks of birds. Thus, increasing radar sensitivity increased the error rate, i.e., the identifications became less specific. The ROC curve

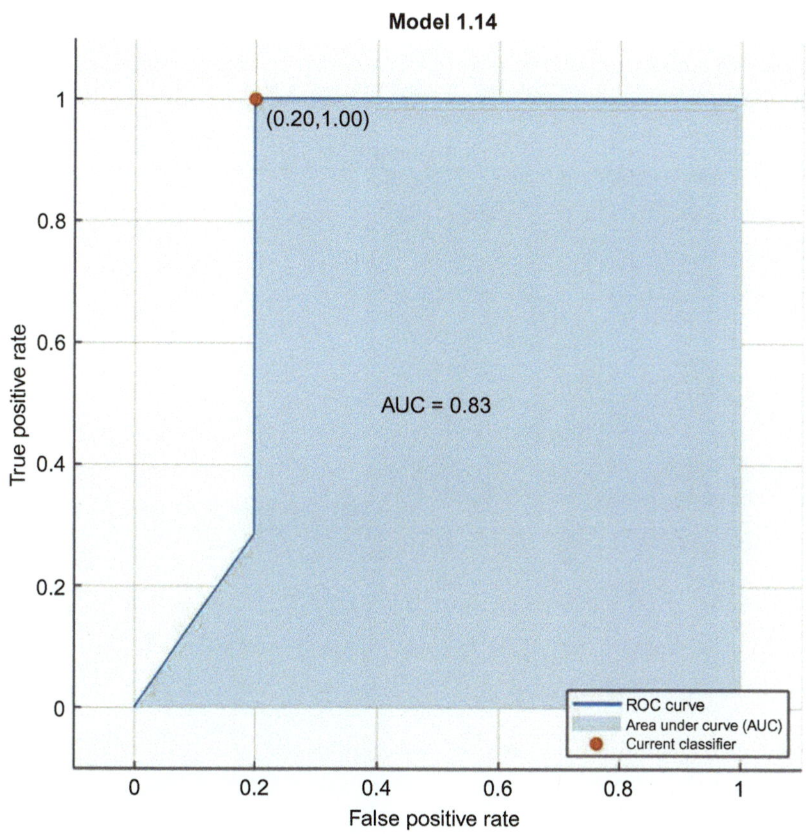

Fig. 3 The ROC and the AUC. The area under the curve (AUV) of the Receiver Operator Characteristic (ROC) is a method to further test the accuracy of the predicative model.

takes this problem into account. A model's true positive identification rate (sensitivity) is measured on the vertical axis, while the model's false positive rate (lack of specificity) is measured on the horizontal axis. In more general statistical terms, specificity is (1—Type 1 error), while sensitivity is (1—Type 2 error). The ROC curve then reflects how well the model trades off sensitivity and specificity, true positives and false positives. The ROC graph measures proportions along the axes. Since it is not possible to get more than 100% correct classifications, the axes are limited to 0,1 and the graph is a unit square. If the ROC is coincident with a 45 degree line in this graph, the model is no better than a coin toss (a coin that is weighted by the proportion of the target variable in the training set) as a predictive tool. The area under such a ROC would be 50% of the unit square. To be better than a coin toss, then any predictor model must have more than 50% of the unit square under the ROC. A model with an AUC of 100% would be a perfect predictor. Even for our small dataset, this measure suggests that the algorithm has an 83% accuracy. What does this mean? For that, we need to delve into Fig. 4, the confusion matrix.

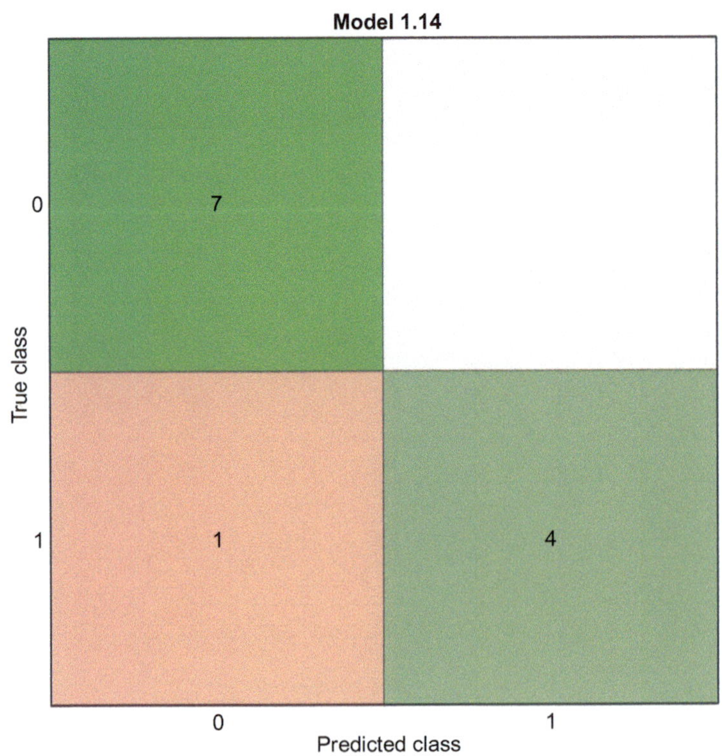

Fig. 4 The confusion matrix. The confusion matrix is used to visually represent how the predictive model preforms on selected test class.

Recall we had 12 patients in our data set. Our algorithm identified the gut microbiome distribution or "profile" that would predict a positive response to KD (reduced seizure incidence) correctly in 4 out of 5 patients (83% accuracy; investigating question 1 and 3 from Table 1). Our algorithm, however, incorrectly identified a patient as someone who did respond to a KD as someone who would not. Nevertheless, our algorithm identified the microbiome distribution that correctly predicted the identity of ALL the patients who did not respond to a KD. This view of model accuracy informs a researcher of different ways of interpreting model accuracy, leaving room for human judgment. What is more important—putting someone on a treatment that they will not respond to or failing to treat someone who would respond to treatment? Machine learning may identify different treatments with different kinds of accuracy in this type 1 and type 2 error sense. Which type of accuracy should have more salience? That is a matter of human judgment. Current machine learning technology cannot replace doctors.

How well does our model do then? We leave that to the researcher's judgment. The user will note that there is no test of significance here. In fact, this is a strength of the machine learning approach. Machine learning militates against (if not completely eliminates) researcher bias. Researchers often use *p*-values to say that a hypothesis is correct. The American Statistical Association, however, has noted that this interpretation is misleading. A *p*-value of 0.01 merely suggests that if the null hypothesis is correct, then there is a 1% chance of finding another result at least as extreme as the observed result. Thus, for example, if one runs a parametric regression 10,000 times (easy enough to do with current computers), then with a *p*-value of 0.01, they will get a "significant" *p*-value 100 times. Thus, the researcher might run the regression 100 times and get significance 100 times out of sheer chance and get a false sense of confidence in their results; this can be dangerous in a clinical setting. Testing out of sample predictive salience can thus be an initial test for whether a particular set of variables matter or not. A significant predictor that does not do a good job of predicting the outcome out-of-sample should come under suspicion, even if the *p*-value is 0.01. Thus, validated machine-learned predictions can provide greater confidence to parametric estimates. Overall, each type of algorithm has its own set of insights. For example, a simple CART tree is easy to interpret since it follows a decision tree pattern; random forest approaches highlight how different predictors separately affect a target; while artificial neural networks (ANNs), if predictively salient, suggest variables act together to influence the outcome in complex networks, rather than individually.

The former indicates that changing individual variables may change the outcome, while the latter suggests a more systemic approach. These applications are extremely relevant from a gut microbiome perspective (and biological research in general), as gut microbiota form complex networks of variable structure and function.

3.3 Identifying the most important predictors

In this next step, we identify the most predictively salient variables (question 2 in Table 1). The process itself is quite simple. The algorithm permutes the input variables and then notes the increase in the misclassification rate.[122] If there is no increase in the misclassification rate, then clearly, the variable is irrelevant. If a variable does not predict, then it cannot be causal. However, this process also allows the algorithm to rank variables in order of predictive importance. For example, in random forests, the algorithm uses variables to separate the unit of analysis (patients, in our case) into purer "bins" by developing a decision tree within a sub-sample of the training data set. The algorithm then records the prediction error of this tree relative to the out-of-bag portion of the data. The classification error rate is used when the target variable can be separated by class, while the mean squared error (MSE) or mean squared deviation (MSD) is used when the target is a continuous variable. It repeats this process for each permutation of the predictor variables. The prediction/classification error is averaged across all trees and normalized. Other methodologies for identifying variable importance rankings are also available, for example, measuring how well a variable splits the unit of analysis into homogenous "bins" (the total decrease in node impurity) averaged across all trees.[123] Fig. 5 shows variable ranking within our data. Variable 1 is the highest-ranked variable. These variables, in our case, are specific gut bacteria, however, these identities were retracted for reasons mentioned earlier in the chapter. The more top-ranked variables deserve scrutiny from a treatment (or causal) point of view. If wet experiments using this more parsimonious list of variables are successful in pinpointing causality, then they become candidates for effective patient-centric research or treatment. The very salience of predictive power should then increase confidence in the success rate of such a treatment. Accurate prediction by its very definition implies that the effect of a change in a now causal variable will be more effective than a variable that, even if causal, is not a very good predictor. It is this part of the workflow that is most effective at speeding up the discovery process. Less effort, time, and money need to be put into lab experiments

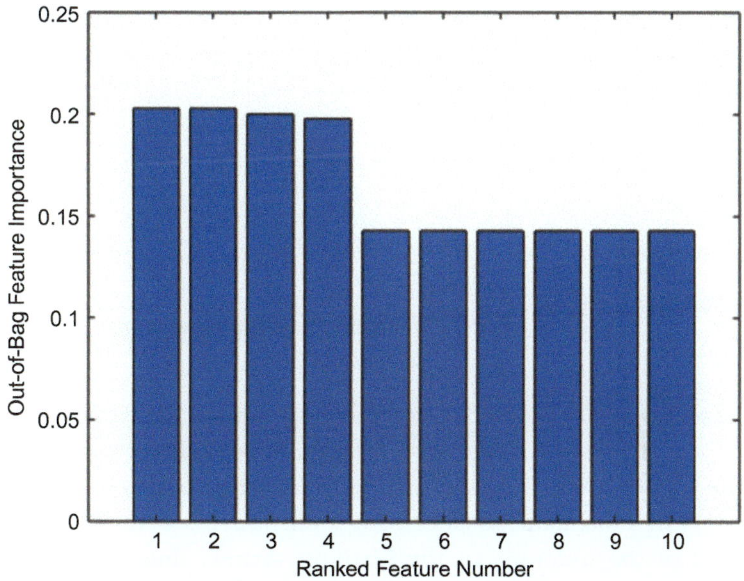

Fig. 5 Variable importance. Predictor importance plot illustrates the most important variables that influence the prediction.

since the researcher can now focus on a more frugal list of potential causal variables—the most important predictors—or most important organisms. That is, the variables at the bottom of the variable importance list are unlikely candidates for manipulation as a treatment. Therefore, experiments to test for causality need not be designed around them. In this particular example, we could expect that the patients with the gut microbiome profile or distribution that mimics the most predictive distributions to be the most positively affected by KD interventions. Such conclusions should only be drawn if our results here are confirmed using much larger datasets and causal links are established. Fair enough, but will patients be helped or hurt by this effect?

3.4 Evaluating the predictors using partial dependence plots

Once we have established that specific predictors are more important than others in predicting an outcome, then, to generate precise hypotheses, we also need to know the direction of this prediction. Does increasing Variable 1 increase or decrease the likelihood of seizures? Partial dependence plots (PDPs) identify the course of these relationships, therefore providing precise causal hypotheses.

PDPs display the marginal effect of variable x_k conditional on the observed values of all other variables, $(x_{1,-k}, x_{2,-k}, \ldots x_{n,-k})$. Specifically, they plot the function:

$$\hat{f}(x) = \frac{1}{n}\sum_{i=1}^{n} f(x_k, x_{i,-k}),$$

where the summand, $f(x_k, x_{i,-k})$, is defined by:

$$f(x) = ln\,[p_1(x)] - \frac{1}{C}\sum_{s=1}^{C} ln\,[p_s(x)].$$

Here, s indexes the classes of the target variable, and C is the total number of classes. So, if an increase in a given factor, x_k, increases the probability of the outcome ($s = 1$), then the value of the function plotted by the PDP also increases. The plots help visualize how a change in a predictor affects the likelihood of the outcome in the learning sample, thus helping identify the marginal effects of the most important variables. Recall that using the PDP in the training set is validated by the lower prediction error in the test set. Fig. 6 reports the PDP for Variable 1. Note that an increase in Variable 1 reduces the likelihood of seizures from slightly above 0.43 to somewhat

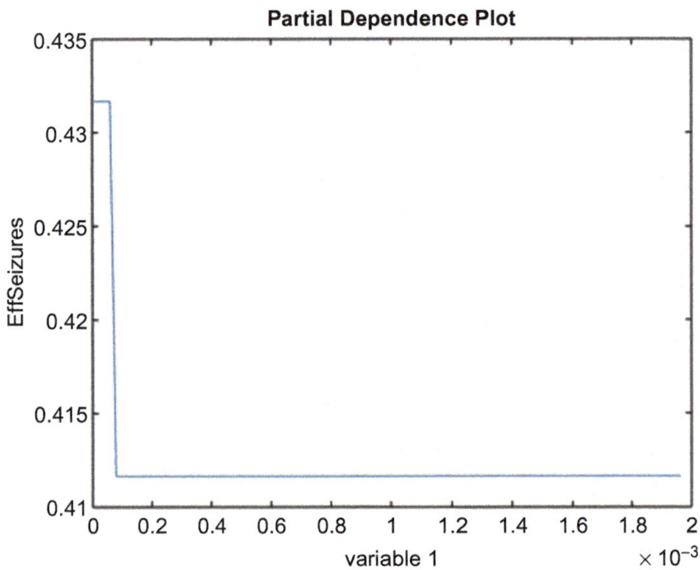

Fig. 6 Partial dependency plots, actionable results. The partial dependency plot focuses on an individual predictor to understand how much the variable affects the outcome.

above 0.41. Is this a meaningful reduction? That is, once again, a matter of judgment for the researcher. Nevertheless, if the researcher judges this to be a meaningful reduction, then one potential hypothesis may be to say that an increase in Variable 1 should reduce the likelihood of seizures. A more precise hypothesis is possible. This effect should only be observable in the 0–0.2 range for Variable 1. Such a hypothesis-generating precision makes it possible to develop a more calibrated experiment design, further reducing time, effort, and money spent on lab experiments.

3.5 Conclusions

Above, we have made a case that using ML techniques can speed up the discovery process, mainly by eliminating variables that are not predictive and, therefore, cannot be causal. But what variables should be included? These need to be driven by what we know about biology, sociology, and economics, i.e., theory matters. Unfortunately, such an initial choice may itself be subject to bias. However, predictive algorithms have a built-in way of reducing such specification biases. Variables chosen out of pure bias (that is to say, an inaccurate "theory" or model) will automatically be eliminated because, by definition, they will not be able to predict well. As an extreme example, an algorithm would rapidly remove leeches as a cure for cancer. Moreover, eliminating non-predictive variables before designing large-scale wet-lab experiments or field trials reduces the risk that a researcher will go through the entire "gold standard" process only to find nothing. Since negative results are also hard to publish, career-minded researchers will find machine learning useful in identifying potential causal pathways that are more likely to lead to positive and publishable results.

4. Future outlook

Machine learning is data driven, thus algorithms with more data to train on will be better at prediction. First, we emphasize that researchers should approach the data collection process by considering the downstream output, collecting as much data as possible. Second, researchers must think about data curation. For example, from a business perspective, firms often collect data that cannot be used in any meaningful way simply because they are stored in ways that require expensive data curation. Last, we have emphasized above that causal machine learning models can and should be developed as a culminating outcome. Nevertheless, this outcome has a workflow of its own. Teams (for example, some of the co-authors of this

Table 3 Future approaches for evaluating gut microbiome dysbiosis and associated diseases.

(1) Longitudinal and interventional studies in well-defined human cohorts identifying microbial candidates through sequencing
(2) Investigating mechanisms by multi-omics which will identify meaningful candidates by analyzing taxonomy as well as genes and pathways of importance through, e.g., shotgun metagenomics or metatranscriptomics sequencing combined with proteomics and untargeted metabolomics
(3) *In vitro* validation of candidates in microbiology labs, e.g., are certain disease-associated metabolites produced by the selected community of candidates or even single species in culture?
(4) Directed manipulation of microbiota in animal studies, e.g., colonizing germ-free or antibiotics-treated mice with relevant single candidates or whole communities and study their effect related to the disease of interest

chapter) are working on further developing this workflow—but there may be multiple ways to use machine learning to bell the causality cat. At the same time, scientists must understand the true nature and limits of causality within machine learning systems.[124]

The gut microbiota is implicated in a wide variety of diseases from inflammatory bowel disease (IBD) and colorectal cancer (CRC), to diseases in the periphery that previously were not associated with gastrointestinal disturbances, e.g., neurological disorders, such as autism spectrum disorder, Parkinson's, or various liver and skin diseases. Most findings, however, are from studies merely describing an association between a "dysbiotic" taxonomic composition and a certain disease. In these descriptive studies, mainly based on 16S rRNA gene sequencing, it is impossible to evaluate whether dysbiosis is a cause or an effect of the disease, as well as what consequences on disease progression they may pose. Scientists need to focus on critically evaluating these associations. This could be done by the following approaches, ideally combining all of them, described in Table 3.

4.1 Characterizing gut microbial functions

Research has highlighted a widespread functional redundancy among bacteria, meaning that many different, not closely related taxa provide similar functions, such as production of short-chain fatty acids (SCFA). Therefore, when considering mechanisms, it is important to analyze the functional differences between the microbiota in different groups within a study, as seemingly spurious differences in several taxa might represent a shift in SCFA production or other functions important for human health. Functional profiling can be based on different data sets, i.e., through interference of functions on whole-shotgun metagenomics sequencing data by gene annotation, analysis of whole community mRNA (metatranscriptomics), analysis of the whole protein pool within a sample (metaproteomics), or untargeted identification and (semi-)quantification of the metabolites in each sample (metabolomics). Again, as all these levels from DNA to metabolites are regulated through various mechanisms, it is ideal to combine several (if not all) of these methods to identify pathways that have the potential to be manipulated for the purpose of maintaining or regaining health. Such integrative analysis, however, is still in its infancy and the field currently lacks appropriate biostatistical tools and standards to correctly integrate these data sets. Metabolomics, for example, can be used to analyze functional differences. Characterizing metabolites and their interactions within the host has been an emerging tool used for precision medicine. Measuring glucose metabolites is an example of how metabolomics has been used to diagnose disease, such as diabetes. We study metabolomics to understand the body's physiological response to disease, along with identifying possible disease indicators.[125] Parallel to current bioinformatics challenges, biases are introduced due to varying sample collection techniques, platforms, and procedures used when analyzing metabolomics data. Metabolomics labs have unique methods which result in varying metabolite coverage and metabolite detection. Thus, a standard operating system is in need in order to obtain metabolomics data, as well as data for the approaches mentioned above, that can be comparable between laboratories.[125]

4.2 Observing the whole picture: Microbial dark matter

As mentioned earlier in the chapter, the bacteriome is the most highly researched and characterized aspect of the gut microbiome. However, the full picture encompasses other equally important "*biomes*" including the virome, mycobiome, archaeome, and eukaryome; or viruses, fungi, archaea, protists and helminths, respectively. More recent studies have demonstrated the importance of viral and fungal gut microbiome components to gut homeostasis.[126,127] From a research perspective, analyzing these variable phylogenies has traditionally been approached through specific and specialized methods, leading to inherent fragmentation in the way we study and understand the gut microbiome ecosystem. Additionally, the gut virome and mycobiome remain largely understudied due to a lack of optimized tools and curated reference databases for classification and identification—leading to potentially unidentifiable, unaligned reads in samples referred to as microbial dark matter.[128129,130]

With the arrival of HTS technologies, particularly whole-shotgun metagenomics, researchers have been empowered to explore the contents of microbial dark matter. Mounting evidence suggests the gut virome is mostly comprised of RNA and DNA bacteriophages, potentially out numbering bacteria 10:1, and playing a critical role in shaping bacterial diversity, horizontal gene transfer, nutrient turnover, and co-evolution.[131,132] The mycobiome tends to encompass species such as *Aspergillus, Candida, Fusarium, Saccharomyces*, and *Penicillium* and the archaeome is dominated by methanogens.[133,134] Developing new methods to explore each of these distinct ecological populations, not only from a host-interaction perspective but also microbial intra-actions, will be essential for further applications in a clinical setting.

4.3 Further applications and considerations

While discussing the effects of diet, particularly the KD, we can expand workflows and considerations like those outlined in this chapter to other disease states, such as the effects of diet on diabetes and the effect of diabetes medications on gut microbiome.[48,135] Indeed, some diabetes medications are known to cause weight loss, such as Metformin, Dapagliflozin, Canagliflozin, Semaglutide, and Dulaglutide. Of these, the SGLT2 group of drugs can cause ketosis, which is thought to contribute to the weight loss observed with these classes of medications. Studies have elucidated the roles of metformin, glucagon-like peptide-1 receptor agonists, inhibitors of dipeptidyl peptidase-4, sodium/glucose cotransporter inhibitors, and other less studied medications on gut microbiota. Antidiabetic drugs mainly effect the gut microbiome composition, appearing to increase the SCFA-producing bacteria responsible for weight loss and inflammation suppression, decrease the *Firmicutes/Bacteroidetes* ratio, and increase the overall efficiency of carbohydrate fermentation. However, the variations in gut microbiota composition and their influence in medicinal efficacy is still poorly understood.[136–138] By applying modified versions of our pipeline to study the intersection of diabetes and the gut microbiome, there's potential to elucidate and/or rule-out certain connections in the future.

4.4 Pharmaceutical and industry outlook

As integrated approaches continue to further our understanding of the gut microbiome's role in human health, interventions that modulate the gut microbiome for a particular outcome have and will continue to be an industry focus. Probiotics and prebiotics, for example, have flooded the market in the last few years, as they are thought to favorably manipulate gut microbial composition and promote health. Probiotics aim to help colonize the gut with beneficial bacteria, such as *Lactobacillus* and *Bifidobacterium* strains associated with a "healthy" gut microbiome. Their delivery methods vary, from encapsulated supplements to yogurt and fermented foods.[139,140] Prebiotics are delivered in similar forms to probiotics, but rather than depositing specific bacterial taxa into the gut (as with probiotics), they "feed" and support the beneficial microorganisms that already exist in the gut; thought to lead to healthy outcomes for the host.[141] Next-generation probiotics is an emerging area of research that examines health-promoting bacteria while targeting specific diseases.[142] Though probiotic and prebiotic usage has been popularized, many clinical conflicts still exist, as much of the current data is still preliminary. Continued research in this area is necessary for promising developments in next-generation probiotics.[143] Similar considerations are applied to our understanding of how the gut microbiome affects drug response and vice versa,[144–147] as recent research has highlighted that therapeutic effectiveness can be variable depending on an individuals' gut microbiota.

Acknowledgments

Charles H King (http://orcid.org/0000-0003-1409-4549), Jonathon Keeney (http://orcid.org/0000-0002-7665-6204), and Juige (April) Yang (http://0000-0002-6497-0714) provided valuable contributions. This project was supported in part by the McCormick Genomic and Proteomic Center (MGPC) at the George Washington University. Stefanie Prast-Nielsen (https://orcid.org/0000-0001-5877-7988) was supported by the Swedish Brain Foundation, Margarethahemmet Society and Stiftelsen Sunnerdahls Handikappfond.

References

1. Madigan MT. *Brock Biology of Microorganisms*. 13th ed. San Francisco: Benjamin Cummings; 2012.
2. Shulman ST, Friedmann HC, Sims RH. Theodor Escherich: the first pediatric infectious diseases physician? *Clin Infect Dis*. 2007;45(8):1025–1029.
3. Podolsky SH. Metchnikoff and the microbiome. *Lancet*. 2012;380(9856):1810–1811.
4. Hug LA, Baker BJ, Anantharaman K, et al. A new view of the tree of life. *Nat Microbiol*. 2016;1:16048.

5. Wade W. Unculturable bacteria—the uncharacterized organisms that cause oral infections. *J R Soc Med.* 2002;95(2):81–83.
6. Steen AD, Crits-Christoph A, Carini P, et al. High proportions of bacteria and archaea across most biomes remain uncultured. *ISME J.* 2019;13(12):3126–3130.
7. Razumov AS. The direct method of calculation of bacteria in water: comparison with the Koch method. *Mikrobiologija.* 1932;1:131–146.
8. Xu HS, Roberts N, Singleton FL, Attwell RW, Grimes DJ, Colwell RR. Survival and viability of nonculturable Escherichia coli and Vibrio cholerae in the estuarine and marine environment. *Microb Ecol.* 1982;8(4):313–323.
9. Reuter JA, Spacek DV, Snyder MP. High-throughput sequencing technologies. *Mol Cell.* 2015;58(4):586–597.
10. Cao Y, Fanning S, Proos S, Jordan K, Srikumar S. A review on the applications of next generation sequencing technologies as applied to food-related microbiome studies. *Front Microbiol.* 2017;8:1829.
11. Zhu S, Jiang Y, Xu K, et al. The progress of gut microbiome research related to brain disorders. *J Neuroinflammation.* 2020;17(1):25.
12. Turnbaugh PJ, Ley RE, Hamady M, Fraser-Liggett CM, Knight R, Gordon JI. The human microbiome project. *Nature.* 2007;449(7164):804–810.
13. Integrative HMPRNC. The integrative human microbiome project. *Nature.* 2019;569 (7758):641–648.
14. Lloyd-Price J, Mahurkar A, Rahnavard G, et al. Strains, functions and dynamics in the expanded Human Microbiome Project. *Nature.* 2017;550(7674):61–66.
15. McDonald D, Hyde E, Debelius JW, et al. American gut: an open platform for citizen science microbiome research. *mSystems.* 2018;3(3).
16. Gilbert JA, Jansson JK, Knight R. The Earth Microbiome project: successes and aspirations. *BMC Biol.* 2014;12:69.
17. Johnson-King B, Terry SF. Future of microbiomes through the national microbiome initiative. *Genet Test Mol Biomarkers.* 2016;20(10):561–562.
18. Team NIHHMPA. A review of 10 years of human microbiome research activities at the US National Institutes of Health, Fiscal Years 2007-2016. *Microbiome.* 2019;7(1):31.
19. Integrative HMPRNC. The Integrative Human Microbiome Project: dynamic analysis of microbiome-host omics profiles during periods of human health and disease. *Cell Host Microbe.* 2014;16(3):276–289.
20. Belizario JE, Faintuch J. Microbiome and gut dysbiosis. *Experientia Suppl.* 2018;109: 459–476.
21. Carding S, Verbeke K, Vipond DT, Corfe BM, Owen LJ. Dysbiosis of the gut microbiota in disease. *Microb Ecol Health Dis.* 2015;26:26191.
22. Shaffer M, Armstrong AJS, Phelan VV, Reisdorph N, Lozupone CA. Microbiome and metabolome data integration provides insight into health and disease. *Transl Res.* 2017;189:51–64.
23. Rastelli M, Cani PD, Knauf C. The gut microbiome influences host endocrine functions. *Endocr Rev.* 2019;40(5):1271–1284.
24. Belkaid Y, Hand TW. Role of the microbiota in immunity and inflammation. *Cell.* 2014;157(1):121–141.
25. Cryan JF, O'Riordan KJ, Sandhu K, Peterson V, Dinan TG. The gut microbiome in neurological disorders. *Lancet Neurol.* 2020;19(2):179–194.
26. Flint HJ. The impact of nutrition on the human microbiome. *Nutr Rev.* 2012;70 (Suppl. 1):S10–S13.
27. Zhang Y-J, et al. Impacts of gut bacteria on human health and diseases. *Int J Mol Sci.* 2015;16:7493–7519.
28. Lukes J, Stensvold CR, Jirku-Pomajbikova K, Wegener Parfrey L. Are human intestinal eukaryotes beneficial or commensals? *PLoS Pathog.* 2015;11(8):e1005039.

29. Minot S, Sinha R, Chen J, et al. The human gut virome: inter-individual variation and dynamic response to diet. *Genome Res.* 2011;21(10):1616–1625.
30. Kapitan M, Niemiec MJ, Steimle A, Frick JS, Jacobsen ID. Fungi as part of the microbiota and interactions with intestinal bacteria. *Curr Top Microbiol Immunol.* 2019; 422:265–301.
31. Belmok A, de Cena JA, Kyaw CM, Dame-Teixeira N. The oral archaeome: a scoping review. *J Dent Res.* 2020;99:630–643. 22034520910435.
32. Vemuri R, Gundamaraju R, Shastri MD, et al. Gut microbial changes, interactions, and their implications on human lifecycle: an ageing perspective. *Biomed Res Int.* 2018; 2018:4178607.
33. Goldenberg RL, Hauth JC, Andrews WW. Intrauterine infection and preterm delivery. *N Engl J Med.* 2000;342(20):1500–1507.
34. Mohammadkhah AI, Simpson EB, Patterson SG, Ferguson JF. Development of the gut microbiome in children, and lifetime implications for obesity and cardiometabolic disease. *Children.* 2018;5(12).
35. Shao Y, Forster SC, Tsaliki E, et al. Stunted microbiota and opportunistic pathogen colonization in caesarean-section birth. *Nature.* 2019;574(7776):117–121.
36. Mulligan CM, Friedman JE. Maternal modifiers of the infant gut microbiota: metabolic consequences. *J Endocrinol.* 2017;235(1):R1–R12.
37. Chong CYL, Bloomfield FH, O'Sullivan JM. Factors affecting gastrointestinal microbiome development in neonates. *Nutrients.* 2018;10(3):274.
38. Nagpal R, Mainali R, Ahmadi S, et al. Gut microbiome and aging: physiological and mechanistic insights. *Nutr Health Aging.* 2018;4(4):267–285.
39. Tasnim N, Abulizi N, Pither J, Hart MM, Gibson DL. Linking the gut microbial ecosystem with the environment: does gut health depend on where we live? *Front Microbiol.* 2017;8:1935.
40. Singh RK, Chang HW, Yan D, et al. Influence of diet on the gut microbiome and implications for human health. *J Transl Med.* 2017;15(1):73.
41. Goodrich JK, Davenport ER, Clark AG, Ley RE. The relationship between the human genome and microbiome comes into view. *Annu Rev Genet.* 2017;51:413–433.
42. Hullar MA, Fu BC. Diet, the gut microbiome, and epigenetics. *Cancer J.* 2014;20(3): 170–175.
43. Vemuri R, Sylvia KE, Klein SL, et al. The microgenderome revealed: sex differences in bidirectional interactions between the microbiota, hormones, immunity and disease susceptibility. *Semin Immunopathol.* 2019;41(2):265–275.
44. Martinson JNV, Pinkham NV, Peters GW, et al. Rethinking gut microbiome residency and the Enterobacteriaceae in healthy human adults. *ISME J.* 2019;13: 2306–2318.
45. Castaner O, Goday A, Park YM, et al. The gut microbiome profile in obesity: a systematic review. *Int J Endocrinol.* 2018;2018:4095789.
46. Menees S, Chey W. The gut microbiome and irritable bowel syndrome. *F1000Res.* 2018;7.
47. Vivarelli S, Salemi R, Candido S, et al. Gut microbiota and cancer: from pathogenesis to therapy. *Cancer.* 2019;11(1):38.
48. Sharma S, Tripathi P. Gut microbiome and type 2 diabetes: where we are and where to go? *J Nutr Biochem.* 2019;63:101–108.
49. Dahlin M, Prast-Nielsen S. The gut microbiome and epilepsy. *EBioMedicine.* 2019;44:741–746.
50. King CH, Desai H, Sylvetsky AC, et al. Baseline human gut microbiota profile in healthy people and standard reporting template. *PLoS One.* 2019;14(9):e0206484.
51. Simonyan V, Chumakov K, Dingerdissen H, et al. High-performance integrated virtual environment (HIVE): a robust infrastructure for next-generation sequence data analysis. *Database (Oxford).* 2016;2016.

52. Shamsaddini A, Pan Y, Johnson WE, et al. Census-based rapid and accurate metagenome taxonomic profiling. *BMC Genomics*. 2014;15:918.
53. Santana-Quintero L, Dingerdissen H, Thierry-Mieg J, Mazumder R, Simonyan V. HIVE-hexagon: high-performance, parallelized sequence alignment for next-generation sequencing data analysis. *PLoS One*. 2014;9(6):e99033.
54. Lindefeldt M, Eng A, Darban H, et al. The ketogenic diet influences taxonomic and functional composition of the gut microbiota in children with severe epilepsy. *NPJ Biofilms Microb*. 2019;5:5.
55. Basuchoudhary A, Bang JT. Predicting terrorism with machine learning: lessons from "predicting terrorism: a machine learning approach". *Peace Econ Peace Sci Public Policy*. 2018;24(4):20180040.
56. Basuchoudhary A, Bang JT, Sen T, David J. *Predicting Hotspots: Using Machine Learning to Understand Civil Conflict*. Rowman & Littlefield; 2018.
57. Wheless JW. History of the ketogenic diet. *Epilepsia*. 2008;49(Suppl. 8):3–5.
58. Neal EG, Chaffe H, Schwartz RH, et al. The ketogenic diet for the treatment of childhood epilepsy: a randomised controlled trial. *Lancet Neurol*. 2008;7(6):500–506.
59. Winesett SP, Bessone SK, Kossoff EH. The ketogenic diet in pharmacoresistant childhood epilepsy. *Expert Rev Neurother*. 2015;15(6):621–628.
60. Pfeifer HH, Lyczkowski DA, Thiele EA. Low glycemic index treatment: implementation and new insights into efficacy. *Epilepsia*. 2008;49(suppl 8):42–45. https://doi.org/10.1111/j.1528-1167.2008.01832.x.
61. Kim SH, Kang HC, Lee EJ, Lee JS, Kim HD. Low glycemic index treatment in patients with drug-resistant epilepsy. *Brain Dev*. 2017;39(8):687–692. https://doi.org/10.1016/j.braindev.2017.03.027.
62. Rezaei S, Harsini S, Kavoosi M, Badv RS, Mahmoudi M. Efficacy of low glycemic index treatment in epileptic patients: a systematic review. *Acta Neurol Belg*. 2018;118(3):339–349. https://doi.org/10.1007/s13760-018-0881-4.
63. Pfeifer HH, Thiele EA. Low-glycemic-index treatment: a liberalized ketogenic diet for treatment of intractable epilepsy. *Neurology*. 2005;65(11):1810–1812. https://doi.org/10.1212/01.wnl.0000187071.24292.9e.
64. Klein P, Tyrlikova I, Mathews GC. Dietary treatment in adults with refractory epilepsy: a review. *Neurology*. 2014;83(21):1978–1985.
65. Masino SA, Rho JM. Metabolism and epilepsy: ketogenic diets as a homeostatic link. *Brain Res*. 2019;1703:26–30.
66. Rogawski MA, Loscher W, Rho JM. Mechanisms of action of antiseizure drugs and the ketogenic diet. *Cold Spring Harb Perspect Med*. 2016;6(5).
67. De Caro C, Iannone LF, Citraro R, et al. Can we 'seize' the gut microbiota to treat epilepsy? *Neurosci Biobehav Rev*. 2019;107:750–764.
68. Olson CA, Vuong HE, Yano JM, Liang QY, Nusbaum DJ, Hsiao EY. The gut microbiota mediates the anti-seizure effects of the ketogenic diet. *Cell*. 2018;173(7): 1728–1741 e1713.
69. Leinonen R, Sugawara H, Shumway M. International nucleotide sequence database C. The sequence read archive. *Nucleic Acids Res*. 2011;39(Database issue):D19–D21.
70. De Spiegeleer M, De Graeve M, Huysman S, Vanderbeke A, Van Meulebroek L, Vanhaecke L. Impact of storage conditions on the human stool metabolome and lipidome: preserving the most accurate fingerprint. *Anal Chim Acta*. 2020;1108:79–88.
71. Liang Y, Dong T, Chen M, et al. Systematic analysis of impact of sampling regions and storage methods on fecal gut microbiome and metabolome profiles. *mSphere*. 2020;5(1).
72. Allaband C, McDonald D, Vazquez-Baeza Y, et al. Microbiome 101: studying, analyzing, and interpreting gut microbiome data for clinicians. *Clin Gastroenterol Hepatol*. 2019;17(2):218–230.

73. Wang Z, Zolnik CP, Qiu Y, et al. Comparison of fecal collection methods for microbiome and metabolomics studies. *Front Cell Infect Microbiol.* 2018;8:301.
74. Williams GM, Leary SD, Ajami NJ, et al. Gut microbiome analysis by post: evaluation of the optimal method to collect stool samples from infants within a national cohort study. *PLoS One.* 2019;14(6):e0216557.
75. Knight R, Vrbanac A, Taylor BC, et al. Best practices for analysing microbiomes. *Nat Rev Microbiol.* 2018;16:410–422.
76. Arnold JW, Roach J, Azcarate-Peril MA. Emerging technologies for gut microbiome research. *Trends Microbiol.* 2016;24(11):887–901.
77. Huttenhower C, Knight R, Brown CT, et al. Advancing the microbiome research community. *Cell.* 2014;159(2):227–230.
78. Eckburg PB, Bik EM, Bernstein CN, et al. Diversity of the human intestinal microbial flora. *Science.* 2005;308(5728):1635–1638.
79. Varadi L, Luo JL, Hibbs DE, et al. Methods for the detection and identification of pathogenic bacteria: past, present, and future. *Chem Soc Rev.* 2017;46(16):4818–4832.
80. Wang WL, Xu SY, Ren ZG, Tao L, Jiang JW, Zheng SS. Application of metagenomics in the human gut microbiome. *World J Gastroenterol.* 2015;21(3):803–814.
81. Schirmer M, Franzosa EA, Lloyd-Price J, et al. Dynamics of metatranscription in the inflammatory bowel disease gut microbiome. *Nat Microbiol.* 2018;3(3):337–346.
82. Zhang X, Deeke SA, Ning Z, et al. Metaproteomics reveals associations between microbiome and intestinal extracellular vesicle proteins in pediatric inflammatory bowel disease. *Nat Commun.* 2018;9(1):2873.
83. Wilmanski T, Rappaport N, Earls JC, et al. Blood metabolome predicts gut microbiome alpha-diversity in humans. *Nat Biotechnol.* 2019;37(10):1217–1228.
84. Chang Y, Hou F, Pan Z, et al. Optimization of culturomics strategy in human fecal samples. *Front Microbiol.* 2019;10:2891.
85. Harris DM, Hata DJ. Rapid identification of bacteria and Candida using PNA-FISH from blood and peritoneal fluid cultures: a retrospective clinical study. *Ann Clin Microbiol Antimicrob.* 2013;12:2.
86. Tringe SG, Hugenholtz P. A renaissance for the pioneering 16S rRNA gene. *Curr Opin Microbiol.* 2008;11(5):442–446.
87. Vetrovsky T, Baldrian P. The variability of the 16S rRNA gene in bacterial genomes and its consequences for bacterial community analyses. *PLoS One.* 2013;8(2):e57923.
88. O'Donnell JL, Kelly RP, Lowell NC, Port JA. Indexed PCR primers induce template-specific bias in large-scale DNA sequencing studies. *PLoS One.* 2016;11(3):e0148698.
89. Hiergeist A, Reischl U, Priority Program Intestinal Microbiota Consortium/quality assessment participants, Gessner A. Multicenter quality assessment of 16S ribosomal DNA-sequencing for microbiome analyses reveals high inter-center variability. *Int J Med Microbiol.* 2016;306(5):334–342.
90. Brumfield KD, Huq A, Colwell RR, Olds JL, Leddy MB. Microbial resolution of whole genome shotgun and 16S amplicon metagenomic sequencing using publicly available NEON data. *PLoS One.* 2020;15(2):e0228899.
91. van den Bogert B, Boekhorst J, Pirovano W, May A. On the role of bioinformatics and data science in industrial microbiome applications. *Front Genet.* 2019;10:721.
92. Bashiardes S, Zilberman-Schapira G, Elinav E. Use of metatranscriptomics in microbiome research. *Bioinf Biol Insights.* 2016;10:19–25.
93. Knight R, Callewaert C, Marotz C, et al. The microbiome and human biology. *Annu Rev Genomics Hum Genet.* 2017;18:65–86.
94. D'Argenio V. Human microbiome acquisition and bioinformatic challenges in metagenomic studies. *Int J Mol Sci.* 2018;19(2):383.
95. Goodrich JK, Waters JL, Poole AC, et al. Human genetics shape the gut microbiome. *Cell.* 2014;159(4):789–799.

96. MacDonald NJ, Parks DH, Beiko RG. Rapid identification of high-confidence taxonomic assignments for metagenomic data. *Nucleic Acids Res.* 2012;40(14):e111.
97. Gerlach W, Stoye J. Taxonomic classification of metagenomic shotgun sequences with CARMA3. *Nucleic Acids Res.* 2011;39(14):e91.
98. Illeghems K, De Vuyst L, Papalexandratou Z, Weckx S. Phylogenetic analysis of a spontaneous cocoa bean fermentation metagenome reveals new insights into its bacterial and fungal community diversity. *PLoS One.* 2012;7(5):e38040.
99. Brady A, Salzberg SL. Phymm and PhymmBL: metagenomic phylogenetic classification with interpolated Markov models. *Nat Methods.* 2009;6(9):673–676.
100. McHardy AC, Martin HG, Tsirigos A, Hugenholtz P, Rigoutsos I. Accurate phylogenetic classification of variable-length DNA fragments. *Nat Methods.* 2007;4(1):63–72.
101. Diaz NN, Krause L, Goesmann A, Niehaus K, Nattkemper TW. TACOA: taxonomic classification of environmental genomic fragments using a kernelized nearest neighbor approach. *BMC Bioinf.* 2009;10:56.
102. Wood DE, Lu J, Langmead B. Improved metagenomic analysis with Kraken 2. *Genome Biol.* 2019;20(1):257.
103. Truong DT, Franzosa EA, Tickle TL, et al. MetaPhlAn2 for enhanced metagenomic taxonomic profiling. *Nat Methods.* 2015;12(10):902–903.
104. Segata N, Waldron L, Ballarini A, Narasimhan V, Jousson O, Huttenhower C. Metagenomic microbial community profiling using unique clade-specific marker genes. *Nat Methods.* 2012;9(8):811–814.
105. Human Microbiome Jumpstart Reference Strains Consortium, Nelson KE, Weinstock GM, et al. A catalog of reference genomes from the human microbiome. *Science.* 2010;328(5981):994–999.
106. Quince C, Walker AW, Simpson JT, Loman NJ, Segata N. Shotgun metagenomics, from sampling to analysis. *Nat Biotechnol.* 2017;35(9):833–844.
107. Almeida A, Mitchell AL, Boland M, et al. A new genomic blueprint of the human gut microbiota. *Nature.* 2019;568(7753):499–504.
108. Quast C, Pruesse E, Yilmaz P, et al. The SILVA ribosomal RNA gene database project: improved data processing and web-based tools. *Nucleic Acids Res.* 2013;41(Database issue):D590–D596.
109. DeSantis TZ, Hugenholtz P, Larsen N, et al. Greengenes, a chimera-checked 16S rRNA gene database and workbench compatible with ARB. *Appl Environ Microbiol.* 2006;72(7):5069–5072.
110. Cole JR, Wang Q, Fish JA, et al. Ribosomal Database Project: data and tools for high throughput rRNA analysis. *Nucleic Acids Res.* 2014;42(Database issue):D633–D642.
111. Cole JR, Chai B, Farris RJ, et al. The ribosomal database project (RDP-II): sequences and tools for high-throughput rRNA analysis. *Nucleic Acids Res.* 2005;33(Database issue):D294–D296.
112. Forster SC, Kumar N, Anonye BO, et al. A human gut bacterial genome and culture collection for improved metagenomic analyses. *Nat Biotechnol.* 2019;37(2):186–192.
113. Cheng L, Qi C, Zhuang H, Fu T, Zhang X. gutMDisorder: a comprehensive database for dysbiosis of the gut microbiota in disorders and interventions. *Nucleic Acids Res.* 2020;48(D1):D554–D560.
114. Ann C, Gregory OZ, Howell A, Bolduc B, Sullivan MB. The human gut virome database. *bioRxiv.* 2019;655910.
115. Goodacre N, Aljanahi A, Nandakumar S, Mikailov M, Khan AS. A reference viral database (RVDB) to enhance bioinformatics analysis of high-throughput sequencing for novel virus detection. *mSphere.* 2018;3(2):e00069-18.
116. Byrd AL, Deming C, Cassidy SKB, et al. *Staphylococcus aureus* and *Staphylococcus epidermidis* strain diversity underlying pediatric atopic dermatitis. *Sci Transl Med.* 2017; 9(397):eaal4651.

117. Philosophy SEo. *Kant and Hume on Causality*; 2008. https://plato.stanford.edu/entries/kant-hume-causality/#InduNeceConnLawsNatu.
118. Ledolter J. *Data Mining and Business Analytics with R*. John Wiley & Sons; 2013.
119. Basuchoudhary A, Bang JT, Sen T. *Machine-Learning Techniques in Economics: New Tools for Predicting Economic Growth*. Springer; 2017.
120. Grootendorst M. *Validating Your Machine Learning Model: Going Beyond k-Fold Cross-Validation*; 2019. https://towardsdatascience.com/validating-your-machine-learning-model-25b4c8643fb7.
121. MathWorks. *Classification Learner*. https://www.mathworks.com/help/stats/classification learner-app.html.
122. MathWorks. *oobPermutedPredictorImportance*. https://www.mathworks.com/help/stats/classificationbaggedensemble.oobpermutedpredictorimportance.html.
123. RDocumentation. *Importance: Extract Variable Importance Measure*. https://www.rdocumentation.org/packages/randomForest/versions/4.6-14/topics/importance.
124. Judea Pearl DM. *The Book of Why: The New Science of Cause and Effect*. New York: Basic Books; May 15, 2018.
125. Clish CB. Metabolomics: an emerging but powerful tool for precision medicine. *Cold Spring Harb Mol Case Stud*. 2015;1(1):a000588.
126. Mukhopadhya I, Segal JP, Carding SR, Hart AL, Hold GL. The gut virome: the "missing link" between gut bacteria and host immunity? *Therap Adv Gastroenterol*. 2019;12:1756284819836620.
127. Forbes JD, Bernstein CN, Tremlett H, Van Domselaar G, Knox NC. A fungal world: could the gut mycobiome be involved in neurological disease? *Front Microbiol*. 2019;9:3249.
128. Vemuri R, Shankar EM, Chieppa M, Eri R, Kavanagh K. Beyond just bacteria: functional biomes in the gut ecosystem including virome, mycobiome, archaeome and helminths. *Microorganisms*. 2020;8(4):483.
129. King CH, Desai H, Sylvetsky AC, et al. Baseline human gut microbiota profile in healthy people and standard reporting template. *PLoS One*. 2019;14(9):e0206484. https://doi.org/10.1371/journal.pone.0206484.
130. Ren Y. Metagenomic analysis of microbial dark matter in healthy human gut microbiome. *Master's Thesis*. The George Washington University; 2019.
131. Sutton TDS, Hill C. Gut bacteriophage: current understanding and challenges. *Front Endocrinol*. 2019;10:784.
132. von Wintersdorff CJ, Penders J, van Niekerk JM, et al. Dissemination of antimicrobial resistance in microbial ecosystems through horizontal gene transfer. *Front Microbiol*. 2016;7:173.
133. Gaci N, Borrel G, Tottey W, O'Toole PW, Brugere JF. Archaea and the human gut: new beginning of an old story. *World J Gastroenterol*. 2014;20(43):16062–16078.
134. Enaud R, Vandenborght LE, Coron N, et al. The mycobiome: a neglected component in the microbiota-gut-brain axis. *Microorganisms*. 2018;6(1):22.
135. Tilg H, Moschen AR. Microbiota and diabetes: an evolving relationship. *Gut*. 2014;63(9):1513–1521.
136. Xourgia E, Papazafiropoulou A, Papanas N, Melidonis A. Anti-diabetic treatment leads to changes in gut microbiome. *Front Biosci*. 2019;24:688–699.
137. Dore FJ, Domingues CC, Ahmadi N, et al. The synergistic effects of saxagliptin and metformin on CD34+ endothelial progenitor cells in early type 2 diabetes patients: a randomized clinical trial. *Cardiovasc Diabetol*. 2018;17(1):65.
138. Jain R, Awal H, Sen S. Using adult stem cells to monitor endothelial dysfunction in diabetes mellitus. *J Diabetes Complications*. 2020;34:107588.
139. Suez J, Zmora N, Segal E, Elinav E. The pros, cons, and many unknowns of probiotics. *Nat Med*. 2019;25(5):716–729.

140. Davani-Davari D, Negahdaripour M, Karimzadeh I, et al. Prebiotics: definition, types, sources, mechanisms, and clinical applications. *Foods*. 2019;8(3):92.
141. Gibson GR, Hutkins R, Sanders ME, et al. Expert consensus document: the International Scientific Association for Probiotics and Prebiotics (ISAPP) consensus statement on the definition and scope of prebiotics. *Nat Rev Gastroenterol Hepatol*. 2017;14(8):491–502.
142. Satokari R. Modulation of gut microbiota for health by current and next-generation probiotics. *Nutrients*. 2019;11(8):1921.
143. Wieers G, Belkhir L, Enaud R, et al. How probiotics affect the microbiota. *Front Cell Infect Microbiol*. 2019;9:454.
144. Nichols RG, Peters JM, Patterson AD. Interplay between the host, the human microbiome, and drug metabolism. *Hum Genomics*. 2019;13(1):27.
145. Mukherjee S, John S. *Lactulose*. Treasure Island (FL): StatPearls; 2020.
146. Newell J. Portal systemic encephalopathy. *Nurse Pract*. 1984;9(7):26–37.
147. Sahota SS, Bramley PM, Menzies IS. The fermentation of lactulose by colonic bacteria. *J Gen Microbiol*. 1982;128(2):319–325.

Index

Note: Page numbers followed by "*f*" indicate figures and "*t*" indicate tables.

V

W